CW00570640

Guideline no. 61

HACCP in organic agriculture: a practical guide

R. Stanley, C. Knight and F. Bodnár

2009

© Campden BRI 2009
ISBN: 978 0 907503 59 0

Station Road, Chipping Campden, Gloucestershire, GL55 6LD, UK
Tel: +44(0)1386 842000 Fax: +44(0)1386 842100

www.campden.co.uk

Campden BRI
food and drink innovation

HACCP in organic agriculture: a practical guide

Guideline no. 61
2009

Ensure both the safety and organic integrity of your crops and animal products by following a HACCP-based approach

Agriculture is an integral part of the food supply chain and there is increasing interest in all aspects of quality assurance in agriculture, including food safety management. A HACCP-based approach to food safety management can be applied throughout the food chain, from farm to fork, and is widely recognised as an effective and logical means for food safety control that is readily applicable to agriculture, including organic and low input production systems.

After a general introduction and overview of the development of HACCP systems for organic agriculture, this guideline specifically provides examples of HACCP exercises for six organic crop and animal product scenarios. Although only theoretical examples, these have been developed and refined following a series of training sessions held as part of the EU Quality Low-Input Food project. The guideline will be of interest not only to those involved in the production of organic raw materials, but also to those who manufacture and sell organic final products.

ISBN 978-0-907503-59-0

Price per copy:
Members of Campden BRI: £60   Non-members: £90

Contents

order on-line@www.campden.co.uk

Softcover - 100 pages

Postback or faxback order form

Please return by fax or post to Carol Newman, Publications Officer,
Campden BRI, Chipping Campden, Gloucestershire, GL55 6LD, UK

Fax to: +44(0)1386 842100
Tel: +44(0)1386 842048 (direct line)
e-mail: pubs@campden.co.uk

www.campden.co.uk

HACCP in organic agriculture: a practical guide

Campden BRI Guideline No. 61 (2009)

Please send me ____ copies of this document
@ £60 each for members of Campden BRI and £90 each for non-members

Please note that Campden BRI publications are protected by copyright. Distributing photocopies without written permission is prohibited. We offer significant discounts on multiple copies of individual publications - please enquire.

Overseas purchasers please add postage and packing - £3.00 per item to a maximum of £12.00 (EC) or £6.00 per item to a maximum of £24.00 (rest of world)

Surname .... First Name ....Dr/Mr/Mrs/Miss/Ms

Position .... Company ....

Address ....

....

Tel .... Fax ....

e-mail address *(please print clearly)* ....

Payment enclosed for £ .... Please send me an invoice ❑ Our order no. is ....

I wish to pay by Visa/Mastercard/Delta/Switch - Card no ....

Card valid from .... to (expiry date) ....

Card holder's name ....

Signed ....

Payment and dispatch: Members of Campden BRI will be sent an invoice. Non-members are required to make payment with order. Please provide credit card details above, make cheques payable to Campden Technology Ltd or contact the Publications Officer for bank transfer details. Please allow up to 21 days for delivery.

Campden Technology Ltd is a subsidiary of Campden BRI, a company limited by guarantee

Ref: jls5291

PREFACE

Agriculture is an integral part of the food supply chain and there is increasing interest on all aspects of quality assurance in agriculture, including food safety management. A HACCP approach to food safety management can be applied throughout the food chain, from farm to fork. Although a HACCP approach is not a legal requirement in primary production in the European Union, it is widely recognised as an effective and logical means for food safety control that is equally applicable to agriculture including organic and low input production systems.

The focus of this manual is on primary production and the scope covers food safety issues, key product quality issues and organic integrity issues. The HACCP technique is demonstrated by the use of a number of case studies which have been developed as part of an EU integrated research project 'Quality Low Input Food' (QLIF). These are on apples, field vegetables, wheat, eggs, milk and pork meat. The product quality attributes include some that have been identified within other work packages of the QLIF project.

The manual is complementary to several other standards and guidance on food safety and product quality, including Campden BRI Guideline documents 10, and 42, Good Agricultural Practice , organic production standards (e.g. IFOAM, EU Regulation and organic private labels), and food safety management system standards (e.g. Codex and ISO 22000), and other QLIF project outputs. It should be used in combination with other guidance as appropriate.

The guidance presented in this document is based on the joint effort of Chris Knight and Richard Stanley (Campden BRI, UK), Ferko Bodnár (Agro Eco, The Netherlands), and with the assistance of Carlo Leifert (Newcastle University, UK). In the development of the guidance many other experts have been consulted, including other QLIF project partners, and their assistance is gratefully acknowledged. The credit for what has been achieved should be shared by all those mentioned above, whilst any inaccuracies are the sole responsibility of the authors.

R Stanley, C. Knight and F. Bodnár

Acknowledgement

The authors gratefully acknowledge funding from the European Community financial participation under the Sixth Framework Programme for Research, Technological Development and Demonstration Activities, for the Integrated Project QUALITY LOW INPUT FOOD, FP6-FOOD-CT-2003- 506358.

Disclaimer

The views expressed in this publication are the sole responsibility of the author(s) and do not necessarily reflect the views of the European Commission. Neither the European Commission nor any person acting on behalf of the Commission is responsible for any use which might be made of the information contained herein.

CONTENTS

INTRODUCTION

Background

The aim of this manual is to provide specific guidance on establishing a HACCP system for the production of organic raw materials. The guidance given is drawn from Codex (*Basic Food Hygiene Texts, 2003*), Campden BRI Guidelines No.10 (*HACCP in Agriculture: a practical guide,2009*) and No.42 (*HACCP: A Practical Guide, 2009*), and the ISO 22000:2005 standard (*Food safety management systems - Requirements for any organisation in the food chain*). The focus is on farm operations (including site selection and preparation, crop agronomy, animal husbandry, fertiliser and plant protection treatments, harvest, storage and transport to the customer) with special reference to food safety issues and quality attributes specific to organic products including organic integrity.

This manual is to help growers and advisors in organic agriculture throughout Europe. However, it does not represent a pre-made HACCP plan that is universally applicable. In each situation, the process of developing a HACCP plan must take into account the specific circumstances of the farm and production operation.

A HACCP study can be applied throughout the food chain, including primary production and processing and its implementation became a legal requirement for food businesses in the EU on 1st January 2006 under Regulation 852/2004, Hygiene of Foodstuffs. The scope of this regulation does not include primary production; however, it is an expectation that control of food safety hazards is demonstrated in agricultural and horticultural production. In addition, the recently developed ISO 22000:2005 standard on food safety management systems, which includes the application of HACCP principles, is applicable throughout the food chain, including primary production.

An additional consideration for organic producers is to ensure compliance with Regulation 834/2007, which defines the requirements for organic production systems. Regulation 834/2007 lays down detailed rules for organic production. However, products of organic origin are also subject to the general hygiene rules in Regulation 852/2004. In terms of food safety, therefore, there is no difference between organic and conventionally produced food raw materials. The food safety hazards are likely to be the same for similar product types and production activities, although the risks, in terms of the likelihood of their occurrence and severity of the consequence of occurrence, may differ. The guidance given in this document outlines how a food safety management system based on a hazard analysis and control principles approach may be applied to primary production, with special reference to organic apples, field vegetables, breadmaking wheat, eggs, milk and pork meat.

This manual not only considers food safety issues but also product quality attributes. Two groups of product quality attributes are distinguished: those that relate to the product, such as physical and sensory attributes, and product integrity issues, which relate to the way the product is produced as specified in organic production standards. Quality attributes and non compliance with the rules in production standards can be treated as hazards using the same hazard analysis and control principles as used for food safety hazards. However, it is preferable that a clear distinction is made between food safety issues and product quality in a HACCP plan.

HACCP and agriculture

The control of food safety hazards in primary production relies principally on reducing the likelihood of introducing a hazard to the product rather than eliminating or reducing the hazard in the product. The former is controlled by Prerequisite Programmes (PRPs) and the latter at Critical Control Points (CCPs). CCPs are identified in the hazard analysis and are accompanied by critical limits which separate what is safe (acceptable) from unsafe (unacceptable) in terms of the hazard. In general, the severity of the consequence in the case of failure of the control measure at a CCP is greater than at a PRP, and therefore CCPs need more frequent monitoring. This distinction between PRP and CCP is one of the features of ISO 22000.

Prerequisite programmes manage the basic environmental and operating conditions that are necessary for the production of safe and wholesome foods. However, two types of PRP are recognised: PRPs and Operational PRPs. PRPs manage hazards or sources of contamination that may occur at many steps of the process and are not specific to a particular process step. Examples of equivalent terms are Good Agricultural Practice, Good Hygiene Practice, etc. On the other hand, there are instances where hazards that are normally considered to be 'generic', and are thus managed by PRPs, will need to be included in the hazard analysis at specific process steps. These are called Operational PRPs. Unlike CCPs, however, they are not accompanied by critical limits, and the monitoring focuses on whether the PRP is implemented, and not on the hazard itself.

An organisation can focus on having many control measures managed by PRPs and only a few managed at CCPs, or the opposite. In some instances, no CCP can be identified and control of food safety issues is by PRPs. Primary production may be such a case in point, and food safety management is based on a well managed prerequisite programme. Even if there are no CCPs in primary production, a HACCP approach is a logical and structured means of providing a food safety control system that is recognised throughout the food supply chain.

Using the manual

There are two parts to this guideline document. The first section gives a brief description of the stages that need to be considered in sequence to develop a HACCP system in organic crop and animal production. The second section presents worked examples for each of the six selected products. Each example describes typical hazards, including food safety issues, and key product quality attributes, including organic integrity issues. The worked examples demonstrate the application of the HACCP technique in each organic production operation.

The worked examples illustrate a detailed approach that may be applied to each production operation based on the implementation of a full HACCP plan. This approach requires relevant expertise in terms of the production process and the necessary resources to establish and implement a detailed HACCP study. It may be most readily adopted by larger enterprises but is applicable to all sizes of business. A detailed approach allows the most benefit to be derived by a business in terms of identifying and controlling food safety hazards and key food quality attributes.

There are no specific rules for the format of the HACCP plan (presentation is a matter of preference). The plan should clearly outline in a logical sequence the process whereby hazards which are significant for food safety are identified, evaluated and controlled. The HACCP plan should provide sufficient technical detail for the study to be effective. The amount of detail shown in the HACCP plan will depend on the objectives of the study and nature of the operation.

An effective HACCP system in organic production will take time and resources to develop. However, the benefits that can be derived from an effective system include meeting customer requirements, demonstrating conformance to legal and food safety requirements and continuous improvement of the management of the crop or animal production process. A HACCP approach is an effective (and cost-effective), logical and structured means of providing an organic food safety control system. The level of sophistication of the HACCP system will depend on the nature and size of the business. A larger business might reasonably be expected to have a more detailed system than a small or medium sized operation.

Although the focus of the manual is on primary production and on-farm operations, the application of HACCP principles and their implementation as exemplified in this manual are applicable to subsequent stages in the supply chain.

SECTION I: HOW TO SET UP AND CONDUCT AN ORGANIC HACCP STUDY

The HACCP system is based on seven principles (cf. Codex Alimentarius Commission 2003), and when conducting a HACCP study in organic production the seven principles of HACCP may be applied as fourteen stages as shown in Figure 1. These include both essential preparation tasks to enable hazard analysis (the 'Planning' stages 1 to 7 described here) and establishing the HACCP plan (Stages 8 to 14 - the principles).

Figure I. Stages in a HACCP study

Stage 1   Management commitment

Stage 2   Define the terms of reference

Stage 3   Select the HACCP team

Stage 4   Describe the product

Stage 5   Identify the intended use

Stage 6   Construct a flow diagram

Stage 7   Confirmation of flow diagram

Stage 8   List all potential hazards associated with each process step, conduct a hazard analysis and consider any measures to control identified hazards (*Principle 1*)

Stage 9   Determine Critical Control Points (*Principle 2*)

Stage 10   Establish critical limits for each CCP (*Principle 3*)

Stage 11   Establish a monitoring system for each CCP (*Principle 4*)

Stage 12   Establish a corrective action plan (*Principle 5*)

Stage 13   Establish verification procedures (*Principle 6*)

Stage 14   Establish documentation and record keeping (*Principle 7*)

Prerequisite programmes

Within crop or animal production businesses there will be many hazards or sources of contamination that are associated with the basic environmental and operating conditions of the production process. There are few, if any, points in the process where specific action is taken to eliminate or reduce a hazard in the product; control relies on reducing the likelihood of contamination in the first place rather than removing it once the contamination has occurred.

The control of these 'generic' hazards arising from the environment or operating conditions is normally part of good agricultural practice (GAP) or good hygiene practice (GHP). They are a pre-requirement to HACCP and should be in place to underpin the HACCP system.

The term 'prerequisite programme' has become the usual way to describe measures that provide the basic environmental operating conditions that are necessary for the safe production of wholesome foods. The prerequisite programme covers a number of basic areas, namely personnel, equipment, the production environment and materials used. Typical examples include:

- There should be appropriate training procedures for staff.

- Preventive maintenance and calibration procedures should be in place.

- There should be established procedures and schedules for cleaning of equipment.

- An effective pest control programme should be in place.

- There should be appropriate control of glass and other foreign bodies.

- Raw materials and finished products should be stored in appropriate conditions.

Effective prerequisite programmes enable the HACCP system to be focussed on the significant product and process food safety hazards that require specific control measures. By 'screening-out' the general hazards, the identification of true critical control points is made easier and may result in the identification of a relatively small number of critical control points that can be effectively managed and where resources can be targeted.

There may also be instances where hazards that are normally considered 'site-wide' or non-process specific and thus managed by prerequisite programmes, will need to be included in the HACCP plan at specific process steps. The ISO 22000 standard describes this as an operational prerequisite. Some hazards in crop and livestock production may fall into this category.

Operational prerequisites

Good Practice (GAP and GHP) encompasses all aspects of crop and livestock production, and in some operations might assume that all steps in the production process are of equal importance, so that resources cannot be targeted. However, this is often not the case; there are specific points in the process which are more important, and where control is applied and is necessary in order to prevent the occurrence of a hazard. The HACCP process highlights these critical steps, so that resources can be targeted. GAP/GHP in practice encompasses both aspects, including those elements that provide the basic operating conditions of the production process and those which are focussed on specific points in the process, due to their central importance in controlling a named hazard.

If we apply the test that a CCP must be specifically in place to eliminate or reduce a named hazard then there may be few if any clear-cut CCPs in some crop and livestock operations. This does not, however, take account of those steps in hese production operations where control should be specifically applied to prevent the occurrence of a named hazard. It would, therefore, be conceivable to devise HACCP schemes where these steps are cited as CCP's or specific hazard control points due to their central importance in controlling a named hazard. It could also be equally argued that it is a central requirement of GAP/GHP, and therefore within the remit of a prerequisite programme. Whether any of these specific points represent controls within pre-requisite programmes or CCP's in the development of a HACCP plan is probably a matter of opinion, and less important than the underlying process and thinking that goes into developing a food safety management strategy.

A helpful approach is to adopt the concept of

- An operational prerequisite (OpPRP) for controls applied at a step where control is necessary to reduce the likelihood of a named hazard or the severity of its occurrence.

- A CCP at a step where control is necessary to prevent or eliminate a hazard or reduce it to an acceptable level in the product.

It is this approach which is followed in the guidance given in this document. However, it is a matter of opinion as to whether the points which are of central importance in controlling a named hazard are called a CCP or OpPRP.

Prerequisite programmes enable the HACCP to focus on the product and process food safety hazards that require specific control. By eliminating the general hazards, the identification of specific control points (defined as CCPs or OpPRPs) is made easier and may result in the identification of a relatively small number of easily managed steps which are central to the control of the identified hazards.

Prerequisite programmes and OpPRPs will need to be documented and records maintained. This should include evidence of their effectiveness, and if not then the remedial action taken.

1.1 Planning stages to enable hazard analysis

Stage 1: Management commitment

Before the HACCP study begins it is essential that there is full commitment at all levels of management in order that the necessary resources, including relevant personnel, are provided to develop, implement and maintain the HACCP system. Without such commitment there is no point in beginning a study.

To help provide evidence of management commitment some companies have an introductory statement in the HACCP plan or Food Safety/Quality Policy which includes a statement on HACCP.

Stage 2: Define the terms of reference

In order for the study to be developed, implemented and fully effective it is essential that the scope and purpose of the study is outlined clearly at the outset. It is therefore necessary to define factors such as:

- Study objective - for example, a statement as to the purpose of the study such as whether food safety issues and/or product quality aspects are to be considered, and at what point in the production operation food safety/quality is applicable (typically this will be on the finished product as transferred to the next stage in the food chain).

- Product(s) and production details - for example, a statement of the type of commodities, location of the farm or business and production activities to be included in the study.

- Hazards - for example, a statement defining which biological, chemical and physical safety and/or quality hazards are to be managed by the HACCP plan.

- Prerequisite programmes (PRP) that underpin the HACCP study (see Stage 8 Hazards and Controls), including the hazard(s) controlled by the PRP.

The HACCP technique is primarily applicable to issues of product safety associated with biological, chemical and physical hazards. There is also an increasing interest in the application of the HACCP technique to identify hazards and control measures associated with product quality issues. Quality includes both product quality attributes and production attributes, such as those relating to organic integrity. The philosophy inherent in the HACCP technique is equally applicable to food safety and clearly defined quality issues. However, it is recommended that HACCP is focussed on safety issues, and where quality issues are included a clear distinction between safety and quality is shown.

It is important that quality attributes are clearly and precisely defined as this will enable the cause or source of the attribute and associated measures to control it to be accurately determined.

Hazard identification

All the food safety hazards that are reasonably expected to occur in relation to the type of product, type of process and actual production facilities should be identified and recorded in the terms of reference. The identification may be based on the following criteria.

- Experience of the organisation, including historical data.

- Industry norms and relevant guidance.

- External information such as legal and customer food safety requirements.

Where regulatory authorities have established maximum limits, targets or other criteria (e.g. pesticides, mycotoxins), the hazard in question may automatically become relevant to the product.

When identifying the hazards in an organic production operation, consideration should be given to the preceding and following links in the food chain (e.g. production of agronomic inputs, postharvest storage and processing), and to the process operation and equipment, services and utilities, and surroundings. In addition many of the hazards will be the same for organic as they are for conventional products, particularly for food safety issues.

The role of hazard identification is to assess the food safety hazards reasonably expected to occur. In making this assessment the hazards should be evaluated according to the severity of the adverse health effects that can be caused by the hazard and the likelihood of their occurrence (prevalence or frequency of occurrence).

Note that hazards should be considered independently of and before identification of prerequisite programmes.

Stage 3: Select the HACCP team

Depending on the size and nature of the operation, development and implementation of the system should, wherever possible, be undertaken by a team who have adequate knowledge and expertise in order to conduct the study, including knowledge of the organic production process and an understanding of HACCP principles and their application.

It is desirable for a small group of individuals to undertake the study. These individuals should have appropriate training and experience, but may also seek specialist advice from outside the group as necessary. It is feasible for one person to develop the system but this individual should have full understanding of the operation, and should wherever necessary seek specialist support or information, to ensure that the study will be effective.

The individuals involved in the study should be identified; preferably, this should include a statement of their relevant knowledge and expertise.

Stages 4 and 5: Describe the essential product characteristics

A full description of the product(s) under study should be prepared, including defining key parameters which influence the safety and/or quality of the product (these will be used in Stage 8). The essential product characteristics to be considered in an organic study include:

- Description of the product - that is, reference to pertinent aspects such as quality criteria (reference to specifications or contracts may also be relevant)
- Production activities, e.g. reference to organic system or standard which is the basis of the production system
- Storage and transport conditions
- Intended use of the product

The intended use should identify the market and/or use by the customer - for example, whether the product is for marketing for direct consumption or further processing - and should encompass any special considerations relevant to the organic product.

Stages 6 and 7: Construct and confirm the flow diagram

Prior to the hazard analysis beginning, it is necessary to define the production process. This will involve careful examination of the process and operations under study and the production of a flow diagram around which the study can be based.

There are no rules for the format of the flow diagram (presentation is a matter of preference), except that all operational steps (of the production process) should be given in a logical sequence. The flow diagram should provide sufficient technical detail for the study to proceed. The amount of detail shown in the flow diagram in respect of the identification of the steps (including operations and activities) in the process will depend on the objectives of the study and nature of the production operation.

It is important to ensure that the flow diagram is an accurate representation of the production operation as it is the basis on which the hazard analysis is undertaken. In crop and animal production there may not be an opportunity for confirmation due to the time based nature of the production cycle. In this case the personnel involved in the HACCP study must ensure that the flow diagram represents the most likely production options.

1.2 Establishing the HACCP plan

Stage 8 Hazards and controls (Principle 1)

Hazard analysis

Using the flow diagram (Stage 5) as a guide, all the potential hazards, as defined in the scope of the study (Stage 1), that may reasonably be expected to occur at each step should be identified. The consideration should include

- hazards that may be present in materials used (introduced at steps preceding the operation)

- hazards that may be introduced from people, equipment or the environment, and

- hazards that may change (proliferate in some way or survive a step designed to eliminate or reduce them to an acceptable level).

In organic production operations, presence on raw materials and contamination may be the main hazards during the production process, but proliferation and survival cannot be ignored as potential hazards. There should be a deliberate policy to ensure that only realistic hazards are identified.

In practice, the decision process will need to take into account the risk associated with any hazard identification, i.e. the likelihood of the hazard causing an adverse effect, taking into account the likely severity of that effect.

In conducting the hazard analysis, therefore, the following criteria should be taken into consideration.

- The source or cause of the hazard, e.g. where or how it can be introduced into the product.

- The probability of the hazard occurring (likelihood of occurrence).

- The severity of the adverse effects caused by the hazard (food safety or quality).

There are a number of tools that may be used to aid the decision making process based on the evaluation of likelihood and severity of occurrence, including logic table, quadrant graph, and scoring system. Further details on the use of these tools is given in Campden BRI Guideline No. 42.

It is important to ensure that only realistic hazards are identified and that controls which represent reasonable precautions are put in place. Records of the hazard analysis should be maintained.

Selection and assessment of control measures

The next step is to specify what control measures should be applied for each identified hazard. Control measures are actions or activities that are applied to prevent, eliminate or reduce the hazard to an acceptable level. The identification of the source or cause in the hazard analysis will help in deciding on the control measures to be implemented.

For practical purposes control measures may be divided into distinct groups (prerequisites and measures applied at CCPs). Three types of control measures can be recognised, as follows:

Control measure	Description
Prerequisite programmes (PRP)	Activities associated with the process, which manage the basic environment and operating conditions of the facilities and process operation, i.e. hazards that are 'generic' and not specific to a particular process step. The consequence of failure could result in a low risk hazard. They are alternatively referred to as Good Agricultural Practice (GAP), Good Hygiene Practice (GHP), etc.
Operational PRPs (OpPRP)	Activities that are associated with a particular process step, which manage specific significant hazards identified in the hazard analysis, but not otherwise managed by Critical Control Points. Regular checking of the effectiveness of an Operational PRP will be required. A loss of control would result in a low risk food safety hazard but timely corrective action must be taken. There must also be an evaluation of the impact of this loss of control on food safety.
Control Measures applied at CCPs	Actions associated with the product at a particular process step, which are specifically applied to prevent or eliminate a significant hazard or reduce the hazard to an acceptable level. Continuous or "real time" monitoring of the effectiveness of the control will be required. Loss of control is likely to result in a high risk food safety issue and will need immediate corrective action.

In practice most if not all control measures in an organic production operation are PRPs or Operational PRPs. The basis of food safety management in crop or animal production is to minimise the likelihood of the introduction and/or proliferation of hazards, as opposed to eliminating or reducing a hazard already present in the product as marketed. Control measures which are PRPs should be identified in the hazard analysis.

Where appropriate, control measures, including PRPs, need to be underpinned by detailed criteria and/or policies and procedures (see Stage 14) to ensure their effective implementation.

Stage 9: Determine Critical Control Points (Principle 2)

For each hazard identified in Stage 8, determine whether the process step is a Critical Control Point (CCP) or if the hazard is fully controlled by the prerequisite programme, and if so whether the control is an Operational CCP. The identification of CCPs and operational PRPs requires professional judgement and may be aided by the use of a decision tree.

In practice, there are few if any control measures in an organic production operation that will eliminate or significantly reduce the hazard level, that are actions associated with the product. There may, however, be a number of control measures that are applied at a specific step, and where the impact of the control measure on the hazard level and the severity of the consequence in case of failure is high, such that the control may be designated an Operational PRP.

Control points in organic production

In an organic production HACCP plan, control measures should be categorised in the hazard analysis by the need to be managed through the prerequisite programme or categorised as Operational PRPs or CCPs, using a logical approach which includes assessments of:

- *Impact of a control measure on the hazard level or frequency of occurrence*: The higher the impact, the more likely the control measure applied will be an Operational PRP or CCP.

- *The severity of the consequence in case of failure of the control measure*: The more severe it is, the more likely the control measure applied will be an Operational PRP or CCP.

- *The feasibility of and need for the control measure to be monitored*: That is, the ability and need to be monitored in a timely manner to enable immediate corrective actions. In general, control measures applied at a CCP require frequent monitoring to ensure that the hazard is controlled effectively, whereas checking of PRPs is limited to whether the control is implemented.

Use of a decision tree

The identification of CCPs/PRPs may be aided by the use of a decision tree. There are a number of decisions trees in use. The most widely recognised is the example given by Codex. For primary production processes a modification to the Codex decision tree may be used (see Figure 2).

In the decision tree the first question is about PRPs, that is 'is the hazard fully controlled by the prerequisite programme?' If the answer is no then the other questions are answered in turn to determine whether it is a CCP or not. If the answer is yes, then there is no need to proceed further with the decision tree. However, it is necessary to determine whether the control is part of the overall prerequisite programme or an Operational PRP which applies to specific steps. This may be aided by using the approach above which includes assessments with regards to: a) the *impact* of the control, b) the *severity of failure* to control, and c) the *need for monitoring* of the control.

Stage 10: Establish critical limits (Principle 3)

For each CCP, the critical limits for the control measures should be identified. The critical limit is the predetermined value for the control measure applied at each CCP, and is the criterion which separates acceptability from unacceptability (e.g. safe from unsafe). It should represent some measurable related parameter that can be assessed quickly and easily.

In primary production operations, critical limits at a CCP are generally quantifiable values, such as time and temperature, whereas PRPs are not generally accompanied by critical limits, as monitoring focuses on whether the PRP is implemented and complied with, and not on the food safety hazard itself.

Stage 11: Establish a monitoring system (Principle 4)

Monitoring is a planned sequence of observations or measurements of control measures. The monitoring system describes the methods which confirm that all CCPs/OpPRPs are under control. It also produces a record of performance for future use in verification (Stage 12). Monitoring must also be able to detect loss of control at the CCP/OpPRP so that corrective action can be taken to regain control (Stage 11).

The monitoring system should preferably address three issues:

- *how* the monitoring is to be carried out - that is, *what* measurement or observation is being carried out, and what record is taken.
- *when* the monitoring is to be carried out - that is, at what frequency is the measurement or observation carried out
- *who* has responsibility for carrying out the monitoring.

Monitoring systems may be continuous (e.g. recording continuous storage temperatures) or discontinuous (e.g. periodic measurement or observation). CCPs need to be monitored in real time in terms of production process operations to ascertain whether hazards are within specified critical limits and enable timely corrective action. Monitoring of OpPRPs focuses on the implementation of the control and tends to be less frequent and not necessarily in real time.

Monitoring should be supported by record keeping. Monitoring records provide an accurate record of performance.

Stage 12: Establish a corrective action plan (Principle 5)

If, in the process of monitoring (Stage 11), it is found that there is a loss of control, it is important that appropriate action is taken. Corrective actions should aim to bring the production process back under control and deal with non-conforming product where appropriate. Corrective actions should preferably involve the consideration of:

*   *what* went wrong
*   *what* is going to be done with the product if it is judged to be non-conforming - that is, out of specification in terms of safety and/or quality criteria
*   *how* to prevent loss of control from happening again in the future
*   *who* has responsibility for the actions taken.

All lots of product that have been affected by a non-conforming situation shall be held under the control of the organisation until they have been evaluated. Products may be released only after evidence that it is safe to do so, for example demonstration that the control measures are effective or analysis of the product indicates that the hazard levels are acceptable. Any lots that have left the organisation may need to be withdrawn or recalled, by notifying relevant interested parties.

It is important that the action taken is logical and rational and should involve a thorough review to determine what action needs to be taken.

Corrective actions should be documented in the HACCP record keeping. This will provide evidence that control has been re-established and safe food is being produced.

Prerequisite programmes

PRPs have to be checked to ensure that they are working effectively; if found not to be effective, appropriate remedial action has to be taken in the same way that monitoring and corrective action is taken for CCPs. The key point is that checking should be carried out frequently enough to ensure process and product safety; this may be less for a PRP than a CCP. In addition, checking of PRPs tends to focus on whether the control is implemented, and not on the hazard itself.

Stage 13: Establish verification procedures (Principle 6)

Verification procedures are used to demonstrate compliance with the HACCP system - that is, that it is operating correctly and effectively. Verification demonstrates conformance (e.g. with stated procedures) and that the HACCP system and prerequisites are effective (i.e. safety requirements are being met). Verification should, therefore, examine the entire HACCP system, including records.

Verification should aim to answer three questions.

- Am I doing what I say I am doing?

- Does the product meet the defined criteria in respect of product legality, safety and quality?

- Is the HACCP plan up to date?

Examples of verification in primary production include:

- Audits of records and associated procedures. These may be internal by the business and/or external by independent third parties such as customers and verifiers of organic assurance schemes.

- Monitoring customer satisfaction - for example, rejections and/or down-grading

- Product testing - for example, analysis of chemical contaminants and quality attributes.

- Review of the HACCP system.

A periodic review of the HACCP plan should be carried out. For crop production this may be once a year. It is essential that the review should consider any changes which affect the HACCP plan or production process, be these internal or external. In addition, there should be an automatic assessment to determine if a review is required when a change occurs outside the normal review period, such as when new feed or equipment have been introduced.

Examples of changes that may trigger an automatic review are:

Internal
- Change to plant protection or livestock disease control strategies
- Change in the production process and technologies employed, including new crop or animal nutrition and plant protection treatments, equipment, storage facilities
- Change in staff levels and/or responsibilities

External
- Emerging knowledge on food safety hazards
- Anticipated change in customer use/requirements
- Changes in legislation on food safety hazards
- Changes in organic standard protocols

Stage 14: Establish documentation and record keeping (Principle 7)

It is important for the operation to be able to demonstrate that the principles of HACCP have been applied correctly, and that documentation and records have been kept in a way appropriate to the nature and size of the business. Records provide evidence that systems operate as specified.

Examples of production documentation include:

- Documentation of the system (the HACCP plan)
- Operating procedures and management policies
- Codes of Practice, organic production protocols, product specifications.

Examples of production records include:

- Operational records, e.g. feeding, veterinary treatments, crop treatments, cleaning and maintenance, staff training records (relating to Stage 5)
- Monitoring data (relating to Stage 8)
- Corrective actions taken (relating to Stage 9)
- Verification data (relating to Stage 10)

The retention period for records should also be considered and defined.

Figure 2. CCP DECISION TREE

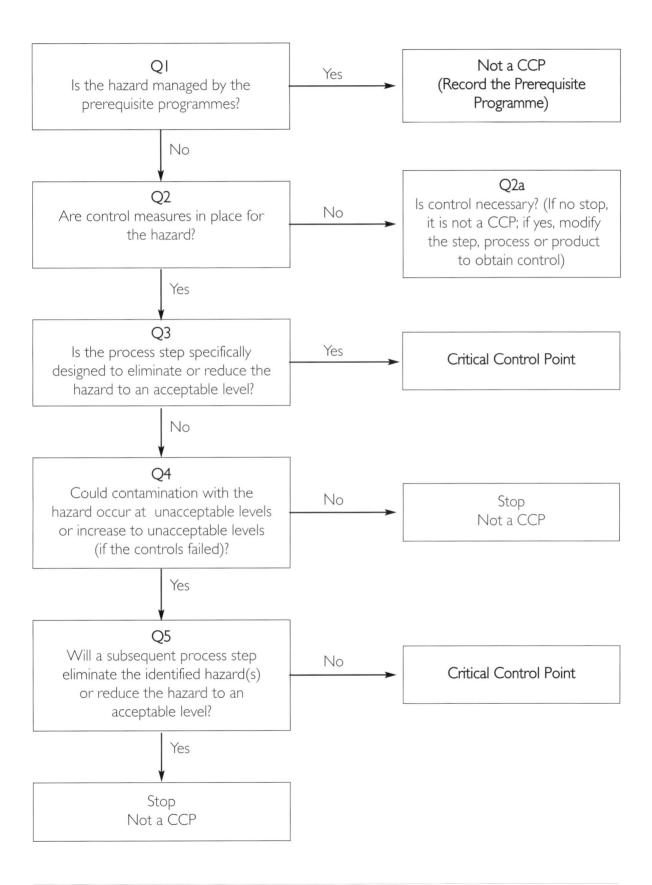

SECTION 2: CASE STUDIES

To demonstrate the application of HACCP principles in organic agriculture, six examples are presented: three each for crop production (apples, field vegetables and bread making wheat) and livestock enterprises (milk, eggs and pork meat). These examples are used to illustrate a detailed approach of the application of HACCP principles to crop production involving the identification of PRPs, operational PRPs and CCPs as applicable. They present typical food safety and key organic quality hazards and controls for a crop and livestock production, which are shown for illustration only and may vary with different crop types and agronomic practices.

The examples are not complete HACCP plans; they are extracts from HACCP plans which focus on the terms of reference, flow diagram and hazard analysis. Some details, such as procedural references for control measures, are not shown. Other details, such as critical limits, monitoring procedures and corrective actions, are shown for indication only and will always depend on specific circumstances.

PRPs are identified at the outset of the HACCP study as they underpin the HACCP plan. The hazards controlled by PRPs and procedures to verify the PRPs are identified separately to the hazard analysis (see the Prerequisite Programme Chart).

Operational PRPs and CCPs (where applicable) are identified as part of the hazard analysis. The hazards and controls for operational PRPs and CCPs are shown in the Hazard Analysis Chart, and include critical limits and monitoring procedures. It should be noted that all the steps in these examples with an identified hazard are operational PRPs. This is because there are no steps in these processes that will eliminate or reduce the hazards to acceptable levels, i.e. they are not considered to be CCPs in these examples. In these examples control of food safety hazards is reliant on minimising the likelihood of introducing hazards and/or their proliferation in the product. However, some operations may wish to highlight important steps in the processes where specific control is applied as a CCP as opposed to an Operational PRP.

For each Operational PRP, monitoring procedures and corrective actions have been established (see the Hazard Analysis Chart). These are shown for indication only and will depend on specific circumstances.

A less sophisticated approach is possible which may be more suited to those organisations with limited HACCP experience and/or resources, particularly small to medium sized businesses. A full HACCP study will, however, provide the greatest flexibility and benefit to a business in terms of identifying and controlling food safety hazards and crop quality aspects at all steps in organic crop and animal production, including storage, and associated activities.

It must be stressed that:

- the details given are for illustrative purposes and are not exhaustive,

- the examples should not be taken as specific recommendations for other production operations,

- the information is not intended for direct use, but only as a demonstration of how the principles of HACCP can be applied to crop and livestock production, and the details given are for guidance only and will depend on specific circumstances.

2.1 Organic breadmaking wheat

2.1.1 Hazards and causes in organic breadmaking wheat

Broadly speaking, hazards associated with primary organic products may be classified into three types: food safety issues, product quality attributes and organic integrity issues relating to the organic status of the product and organic production system. There are numerous theoretical food safety and product quality hazards in primary agricultural products, but a few will only be of significance in a particular agricultural situation, be this organic or conventional agriculture. In deciding the significance of a hazard, the risk associated with the hazard will have to be taken into account.

A food safety hazard is a biological (particularly microbiological), chemical or physical agent in or on a food product with the potential to cause an adverse effect on the health of the consumer. This is very much in line with food safety legislation in Europe, which has the aim of protecting the consumer.

A product quality hazard may be defined as an attribute that causes an adverse effect on the acceptability of the product to the customer, whether this is the consumer or food business using the product. Quality attributes in wheat may relate to physical attributes (e.g. foreign body contamination, Hagberg falling number), composition aspects (moisture content, protein content and quality), and sensory characteristics (e.g. odours).

An organic integrity hazard is defined as the product and production process not complying with the adopted organic standard and production protocol, be this set by legislation, an organic organisation such as IFOAM, or a private organic standard.

The actual hazards in any particular situation, however, will depend on the specific circumstances, including the production system, production location, product type and intended market. Many of the hazards will be the same for organic as conventionally produced products, particularly food safety issues, but the risk of the hazard may be different. In some cases the risk of a hazard may be perceived differently in conventional and organic products, for example the presence of pesticide and veterinary product residues, due to the nature of the respective systems. However, many of the 'extended product' quality attributes which relate to the way the product is produced and the nature of the production system may be unique to organic agriculture.

There are different organic standards for wheat. In Europe, the EU Regulation 834/2007 is the legal minimum requirement that needs to be met for products that are designated 'organic'. Most national implementation rules are very similar to the Regulation. However, private standards may differ and impose a higher standard, including additional requirements. An example of a private standard is the Soil Association in the UK. Demeter is a private standard for biodynamic agriculture that goes beyond organic farming. Details of the differences between the EU Regulation and private standards in Europe are available on the Organic Rules website *www.organicrules.org*.

Possible causes

It is important to be aware and understand the possible cause or source of the identified hazards, as this will help determine the risk associated with the hazard and the most effective control measures. It should noted that the cause of a hazard and the level of risk may be of similar in both organic and conventional production, for example introduction of food-borne pathogens from people or equipment. However, there may be some causes that are specific to organic production, for example compromising organic integrity, or the level of risk may be perceived to be greater in organic production, for example the introduction of pathogens from organic manure.

To identify the cause of a hazard it is helpful to distinguish between those that are *present* in raw materials (e.g. inputs such as organic manure, irrigation water), those that are *introduced* during the production process (e.g. from or due to people, equipment, the environment), and changes to pre-existing hazards by *growth* or *survival* of the hazard. In crop production, presence and introduction are the predominant causes of hazards, although there may be growth of mycotoxin-producing fungi and introduction of mycotoxins.

When considering the causes of loss of product quality and organic integrity issues, it is necessary to take account of the practices outlined in the adopted organic production standard. For example, introduction of a hazard may be due to not adhering to the requirements as specified in the standard. Some possible causes may apply to both organic and conventional production systems, but the risk and perceptions may be different.

For example,

- In organic farms the risk of Fusarium mycotoxins in cereal crops may be perceived to be greater as the use of pesticide intervention treatments is limited.

- In organic farms the risk of low protein content may be perceived to be greater as there is less opportunity to increase the availability of nitrogen for the crop at the applicable growth stage.

2.1.2 Examples of typical hazards and the cause/source

Hazards

Hazard category	Hazard type	Hazard
Food safety	Biological	Food-borne pathogens: pathogenic bacteria (e.g. *E.coli*, *Salmonella*)
	Chemical	Pesticide residues Mycotoxin residues Food contaminant residues (e.g. heavy metals)
	Physical	Contaminants (e.g. ergots, glass)
Food quality	Biological	
	Chemical	Moisture content Protein content (breadmaking wheat) Hagberg Falling Number (enzyme activity)
	Physical	Specific weight Screenings and admixture Visual contaminants, e.g. insect infestation, ergot, and damaged, sprouted, mouldy, shrivelled, pink, or green grains Sensory characteristics (e.g. off odours) Varietal purity
Organic integrity	Physical	Organic standard requirements, e.g. • genetically modified organisms (GMOs) • plant protection treatments not authorised for use in organic production • fertilisers/soil conditioners not authorised for use in organic production • seed not of organic origin • variety differentiated from non organic production

Causes/sources of typical hazards

Hazard	Cause
Food-borne pathogens	Whether present in inputs, e.g. organic manure at unacceptable levels Whether they can be introduced from • equipment, e.g. vehicle cleanliness • environment, e.g. pests (rodents, birds), stock and domestic animals
Pesticides	Whether substances are permitted for use in organic systems and are used in the recommended manner Whether unacceptable residues are introduced from • inaccurate application due to people, faulty equipment • environmental contamination (e.g. soil, adjacent crops, water) • equipment contamination
Mycotoxin	Whether introduced from growth of fungi on grain pre-harvest (Fusarium mycotoxins) predisposed by previous crop, tillage regime, choice of variety, growing and harvest conditions Whether introduced from growth of fungi on grain post-harvest (Ochratoxin A) including microclimates (temperature and moisture) in store conducive to fungal growth and mycotoxin production
Glass	Whether introduced from the environment, e.g. lights in post-harvest handling and storage structures
Moisture content	Whether grain is dried to an appropriate moisture content post-harvest and before storage or transport Whether grain condition is maintained in store and transport
Protein content	Whether it is influenced by the choice of variety Whether it is influenced by failure to provide adequate crop nutrition (fertility management programme)
Insect infestation	Whether there is increase in population during storage due to poor grain condition (moisture content, temperature) Whether there is contamination from handling equipment, store structures (cleanliness)
Varietal purity	Whether variety admixture is present in seed as sown Whether there is contamination from equipment, storage structures (cleanliness) Whether varieties are mixed post-harvest
Organic integrity	Whether there is compliance with the adopted standard, e.g. • GMOs (e.g. seeds) are not used • seed is of organic origin • authorised plant protection treatments, fertilisers and soil conditioners are used

2.1.3 Example HACCP plan: organic breadmaking wheat

[*Name of Producer*] ORGANIC BREADMAKING WHEAT HACCP PLAN

Last review date
(Stage 13)†   *dd/mm/yy*

Terms of ref. *(Stage 2)*

Objective:	This is a demonstration study covering food safety and key crop quality attributes of organic bread wheat at the point of despatch of the crop to the customer (marketing organisation or miller).
Business:	*Name and location of producer*
Process:	Crop production, harvesting and storage of wheat from selection of site to transport of the grain to customers.
Hazards:	**Microbiological**: Food-borne bacterial pathogens (*E.coli* and *Salmonella*)* **Chemical**: Mycotoxin (Ochratoxin A)* **Physical**: Glass *, protein content, integrity of organic products ** Food safety hazards*
Prerequisite programmes:	The HACCP is underpinned by prerequisite programmes (PRPs) which are selected to maintain a hygienic crop production and post-harvest handling environment. Details of the PRPs, hazards controlled by the PRP and verification procedures are given in the Prerequisite Programme Chart.
HACCP team and skills *(Stage 3)*	Farm Manager (agronomy and management skills), and Agronomist (consultant with agronomy and HACCP skills).
Essential product characteristics *(Stages 4 & 5)*	• Product: organic bread wheat • Grain treatments: Cleaning and continuous heated-air drying • Storage conditions: ≤15% moisture content (mc) and ≤15°C temperature depending on the storage period and market requirements. • Grain handling: Grain is handled and stored in bulk. Different varieties and/or grain for specific markets are kept separate. • Intended use: For marketing as a raw material for human consumption.
Flow diagram *(Stages 6 & 7)*	There are five key steps in the operation. These comprise seed selection, crop agronomy (establishment to maturity), crop harvest, grain storage and grain transport. The sequence of operations and product flow are shown in the Process Flow Diagram. The flow diagram was verified by the team as representing the most likely production options on *dd/mm/yy*.

Hazard analysis *(Stages 8 to 12)*	The hazards and controls are detailed in the Hazard Analysis Chart. In wheat production control of hazards is dependent on reducing the likelihood of the introduction or proliferation of a hazard. No CCPs are identified. For hazards at specific steps the controls are designated operational PRPs (OpPRP), and monitoring procedures and corrective actions have been established.
Verification *(Stage 13)*	The following verification procedures are undertaken: Audits of the HACCP system • Internal by the company of PRPs, operational PRPs - at least annually prior to the HACCP review • External by verifiers of the adopted organic assurance scheme. Monitoring of customer satisfaction, including anomalies and rejections Product testing, e.g. pesticide and mycotoxin residue analysis by the producer or producer's customer or through participation in a third party monitoring system Review of the HACCP system: • Periodic review, e.g. annually before new crop season • Prior to significant changes (outside the periodic review) to the process operation or hazards.
Documents and record keeping *(Stage 14)*	The following documents are retained: • The HACCP plan (including previous versions) • Management policies and operational procedures • Relevant organic scheme protocols, Codes of Practice, guidelines The following records are taken: • Operational records, e.g. site and crop management records, crop input applications • Monitoring data • Corrective actions taken • Verification data Records are retained for a minimum of three years.

† Stages in a HACCP study (see Table 1)

Wheat production flow diagram

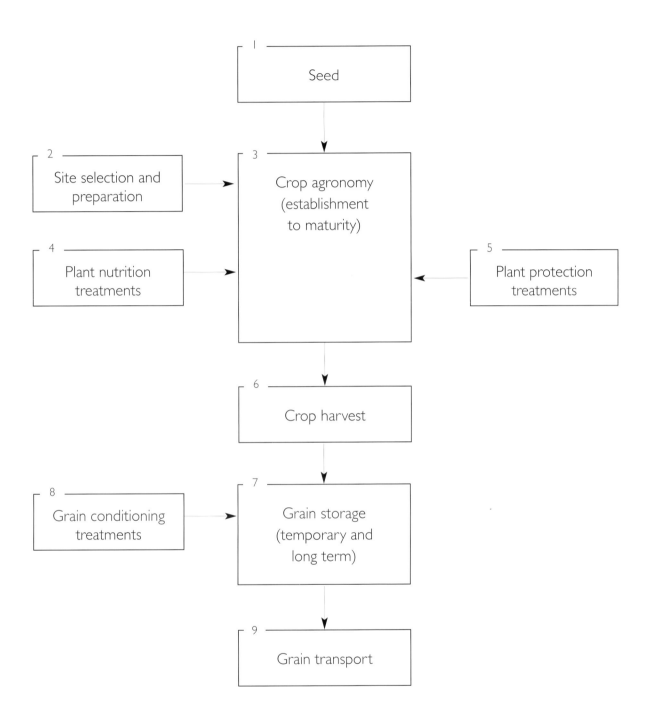

1. Seed

2. Site selection and preparation

3. Crop agronomy (establishment to maturity)

4. Plant nutrition treatments

5. Plant protection treatments

6. Crop harvest

7. Grain storage (temporary and long term)

8. Grain conditioning treatments

9. Grain transport

Prerequisite Programme Chart

Prerequisite programme (PRP)	Hazards controlled by the PRP	Checking procedures	Remedial actions
Personal hygiene Personal hygiene standards: - staff are made aware of the importance of their own personal hygiene, - suitable and sufficient hand washing and toilet facilities are provided, - persons with communicable enteric diseases are excluded from handling food materials.	Introduction of food-borne pathogens from people handling grain.	Scheduled inspection (at an appropriate frequency) of hygiene standards, and record.	Review procedures and training needs. Record actions taken.
Hygiene and housekeeping All equipment and facilities are routinely cleaned to a defined schedule. Includes store structures, equipment, vehicles and trailers used to handle and transport grain on-farm	Introduction of food-borne pathogens from equipment used for harvesting and handling grain.	Scheduled inspection (at an appropriate frequency) of equipment and facilities for fitness for purpose, and record.	Review procedures and take appropriate action to remedy any defects. Record actions taken.
There is a clean-as-you-go policy for spillage.	Introduction of mycotoxin-producing fungi from previous crop residues in store structures	Scheduled inspection before loading of stores for fitness for purpose, and record.	Review procedures and take appropriate action to remedy any defects. Record actions taken.
	Contamination of organic products with non-organic crop residues and other materials from equipment, store structures, etc.	Scheduled inspection (at an appropriate frequency) of equipment and facilities for fitness for purpose, and record.	Review procedures and take appropriate action to remedy any defects. Record actions taken.
Pest control Rodent control procedures - Inspection and treatment of premises to deter and eradicate pest ingress (rodents and birds) - Domestic and stock animals are excluded from grain handling/store areas. - Premises are designed and maintained to prevent entry of birds and rodents	Introduction of food-borne pathogenic bacteria from pests (birds and rodents) and animals (domestic and livestock).	Scheduled inspection (at an appropriate frequency) of facility for pest and animal activity.	Review procedures and take appropriate action to remedy any potential pest ingress. Record actions taken.

Prerequisite programme (PRP)	Hazards controlled by the PRP	Checking procedures	Remedial actions
Glass policy Glass is covered or guarded to prevent contamination of goods. Glass breakage procedures are in place.	Introduction of glass from equipment and the environment (machinery, lights, windows, etc. - grain handling areas).	Scheduled inspection (at an appropriate frequency) of glass fixtures and fittings.	Take appropriate action to remedy any defects. Record actions taken.
Supplier assurance Raw materials (seed, crop nutrition and plant protection treatments, etc.) are sourced from reputable suppliers. Purchasing of materials takes into account the source and the treatement that it may have undergone.	Presence of food-borne pathogens, and of materials not authorised for use in organic production due to contamination at previous stages.	Review approved supplier list. Supplier performance is evaluated seasonally.	Review suitability of suppliers.
Organic protocols The adopted organic standard is complied with in totality	Loss of organic integrity due to use of inputs not authorised by organic protocols (seed, fertilisers/soil conditioners, plant protection treatments)	Planned internal audit of procedures (at an appropriate frequency). Review of external audits by third parties and verifiers of organic schemes.	Review procedures and take necessary actions for any non-compliance identified. Record actions taken
Training Personnel are trained, instructed and supervised commensurate with their activity and competent to undertake the tasks required. Where persons are engaged in activities relating to operational PRPs and CCPs, relevant training is given.	Introduction of hazards or loss of organic integrity due to failure to follow correct procedures.	Scheduled review (at an appropriate frequency) of staff competence to carry out activities.	Review training needs. Record actions taken.
Transport policy Vehicles used for carrying grain off-farm must be clean and fit for purpose Loaded vehicles must be covered.	Introduction of pathogenic bacteria and non-organic materials from vehicles and during transport.	All vehicles collecting grain are inspected before loading for fitness for purpose and covered for transport. The previous three loads are assessed for compatibility.	Take appropriate action to remedy any defects. Record actions taken.

Hazard Analysis Chart

Process step	Hazard	Control	CCP or Op PRP	Critical limit	Monitoring procedure	Corrective action
1. Seed Selection of variety, purchase and receipt of seed	No identified hazard not covered by a PRP		Op PRP			
2. Crop agronomy Crop establishment and management of the developing crop from emergence to maturity.	No identified hazard not covered by a PRP					
3. Site selection and preparation Selection and preparation of land for the wheat crop, including any cultural operations.	No identified hazard not covered by a PRP					
4. Plant nutrition treatments Selection, sourcing and application of fertilisers and soil conditioners pre and post crop establishment.	Introduction of food-borne pathogens from organic manure	Permitted materials are used in the recommended manner (method and timing of application) taking into account the source and treatment that it has undergone prior to use.	Op PRP		Scheduled check of manure use.	Review procedures. Evaluate risk to crop of possible misuse. Record actions taken.
	Failure to attain adequate wheat protein level for breadmaking due to insufficient crop nitrogen supply	Fertility management programme designed to provide breadmaking wheat containing adequate protein	Op PRP		Check available nitrogen levels and grain protein content (to meet customer specification)	Review fertility management programme. Record actions taken

Process step	Hazard	Control	CCP or Op PRP	Critical limit	Monitoring procedure	Corrective action
5. Plant protection treatments Selection, sourcing and application of plant protection treatments.	No identified hazard not covered by a PRP					
6. Crop harvest Harvesting operations, including combining and transfer to storage facilities	No identified hazard not covered by a PRP					
7. Grain storage Temporary storage of grain pre-conditioning, and long term storage of conditioned grain.	Introduction of mycotoxins due to spoilage of unconditioned damp grain	Storage procedures - harvested grain is dried promptly or cooled if drying delayed - high mc grain ($\geq$18% mc) is dried immediately - moist grain (15–18% mc) is dried promptly or cooled ($\leq$15°C) and dried within specified time.	Op PRP		Scheduled check of grain condition (at appropriate frequency)	Review storage procedures Review suitability of grain in store and initiate appropriate stock control, dry and cool as necessary
	Introduction of mycotoxins due to spoilage of conditioned grain (long term store)	Storage procedures - store grain cool ($\leq$15°C) and dry ($\leq$15% mc) to maintain condition	Op PRP		Scheduled check of grain temperature and condition (at appropriate frequency)	Review storage procedures. Review suitability of grain in store and initiate appropriate stock control where necessary
8. Grain conditioning Drying and cleaning of grain.	No identified hazard not covered by a PRP					
9. Grain transport Includes removal from store and loading of transport vehicles.	No identified hazard not covered by a PRP					

mc – moisture content

2.2 Organic field vegetables (lettuce)

2.2.1 Hazards and causes in organic vegetables

Broadly speaking, hazards associated with primary organic products may be classified into three types: food safety issues, product quality attributes and organic integrity issues relating to the organic status of the product and organic production system. There are numerous theoretical food safety and product quality hazards in primary agricultural products, but a few will only be of significance in a particular agricultural situation, be this organic or conventional agriculture. In deciding the significance of a hazard, the risk associated with the hazard will have to be taken into account.

A food safety hazard is a biological (particularly microbiological), chemical or physical agent in or on a food product with the potential to cause an adverse effect on the health of the consumer. This is very much in line with food safety legislation in Europe, which has the aim of protecting the consumer.

A product quality hazard may be defined as an attribute that causes an adverse effect on the acceptability of the product to the customer, whether this is the consumer or food business using the product. Quality attributes in vegetables may relate to physical appearance (e.g. shape, size, colour, defect levels), composition including nutritional aspects, and sensory characteristics including flavour and texture.

An organic integrity hazard is defined as the product and production process not complying with the adopted organic standard and production protocol, be this set by legislation, an organic organisation such as IFOAM, or an private organic standard.

The actual hazards in any particular situation, however, will depend on the specific circumstances, including the production system, production location, product type and intended market. Many of the hazards will be the same for organic as conventionally produced products, particularly food safety issues, but the risk of the hazard may be different. In some cases the risk of a hazard may be perceived differently in conventional and organic products, for example the presence of pesticide and veterinary product residues, due to the nature of the respective systems. However, many of the 'extended product' quality attributes which relate to the way the product is produced and the nature of the production system may be unique to organic agriculture.

Some examples of typical food safety and quality variables in organic vegetables are shown in section 2.2.2.

There are different organic standards for vegetables. In Europe, the EU Regulation 834/2007 is the legal minimum requirement that needs to be met for products that are designated 'organic'. Most national implementation rules are very similar to the Regulation. However, private standards may differ and impose a higher standard, including additional requirements. An example of a private standard is the Soil Association in the UK. Demeter is a private standard for biodynamic agriculture that goes beyond organic farming. Details of the differences between the EU Regulation and private standards in Europe are available on the Organic Rules website *www.organicrules.org*.

Possible causes

It is important to be aware and understand the possible cause or source of the identified hazards, as this will help determine the risk associated with the hazard and the most effective control measures. It should noted that the cause of a hazard and the level of risk may be of similar in both organic and conventional production, for example introduction of food-borne pathogens from people or equipment. However, there may be some causes that are specific to organic production, for example compromising organic integrity, or the level of risk may be perceived to be greater in organic production, for example the introduction of pathogens from organic manure.

To identify the cause of a hazard it is helpful to distinguish between those that are *present* in raw materials (e.g. inputs such as organic manure, irrigation water), those that are *introduced* during the production process (e.g. from or due to people, equipment, the environment), and changes to pre-existing hazards by *growth* or *survival* of the hazard. In crop production, presence and introduction are the predominant causes of hazards.

When considering the causes of loss of product quality and organic integrity issues, it is necessary to take account of the practices outlined in the adopted organic production standard. For example, introduction of a hazard may be due to not adhering to the requirements as specified in the standard. Some possible causes may apply to both organic and conventional production systems, but the risk and perceptions may be different. For example,

- In organic farms the risk of pathogenic bacteria in ready-to-eat vegetables may be perceived to be greater as there is greater use of organic manure.

In organic farms the risk of pesticide residues even at low levels may be perceived to be greater.

2.2.2 Examples of typical hazards and the cause/source

Hazards

Hazard category	Hazard type	Hazard
Food safety	Biological	Food-borne pathogens: pathogenic bacteria (e.g. *E.coli*, *Salmonella*), viruses and protozoa
	Chemical	Pesticide residues
		Food contaminant residues (e.g. heavy metals, nitrates)
	Physical	Foreign bodies (e.g. glass, metal, wood)
Food quality	Biological	Spoilage organisms (e.g. fungal and bacterial rots)
	Chemical	Nutritional attributes (e.g. vitamins, antioxidants)
	Physical	Defects (e.g. pest and disease damage)
		Sensory characteristics (e.g. colour, flavour and texture)
Organic integrity	Physical	Organic standard requirements, e.g. • permitted seed for organic crops • crop variety suitable for organic crops • site suitable for organic crops • permitted pesticides for organic crops • permitted fertiliser and manure for organic crops

Cause/source

Hazard	Cause
Food-borne pathogens	Whether introduced from • people (personal hygiene) • unclean equipment • environment (e.g. pests, stock and domestic animals) Whether present in inputs (e.g. irrigation water, organic manure)
Pesticides	Whether substances are permitted in organic systems and are used in the recommended manner Whether introduced from • inaccurate application due to people, equipment • environmental contamination (e.g. soil, adjacent crops) • equipment contamination
Glass	Whether introduced from • people • environment (e.g. lights)
Seed	Whether permitted seed from recognised source is used
Crop variety	Whether varieties grown in parallel production are different and easily differentiated.
Site	Whether organic production takes place on clearly defined units of land such that production and post-harvest handling areas are clearly separated from non-organic production areas
Manure	Whether prohibited materials are used. Whether manure management and application compiles with the adopted organic protocol
Mineral fertilisers and supplementary plant nutrients	Whether permitted materials are used in the recommended manner.

2.2.3 Example HACCP study

[*Name of Producer*] ORGANIC FIELD VEGETABLE HACCP PLAN (LETTUCE)

Last review date
(Stage 13)†   *dd/mm/yy*

Terms of ref. *(Stage 2)*

Objective:	This is a demonstration study covering organic integrity criteria and food safety for lettuce at the point of despatch of the crop to the customer (produce marketing organisation or retailer).
Business:	*Name and location of producer*
Process:	Crop production, harvesting and storage of lettuce from selection of site to transport of the crop product to customers.
Hazards:	**Microbiological**: Food-borne bacterial pathogens (*E.coli* and *Salmonella*)* **Chemical**: Pesticide residues above statutory MRLs *, Non-permitted pesticide products **Physical**: Glass*, integrity of organic products * Food safety hazards
Prerequisite programmes:	The HACCP is underpinned by prerequisite programmes (PRPs) which are selected to maintain a hygienic crop production and post-harvest handling environment. Details of the PRPs, hazards controlled by the PRP and verification procedures are given in the Prerequisite Programme Chart.
HACCP team and skills *(Stage 3)*	Farm Manager (agronomy and management skills), Agronomist (consultant with agronomy and HACCP skills), Harvest Manager (production skills).
Essential product characteristics *(Stages 4 & 5)* Shelf life: 5 days.	• Product: Whole head lettuces (Crisp, Romaine, etc.) • Storage and transport conditions: ≤5°C. • Packaging; Each lettuce is packed into a polythene bag and the bagged lettuces are placed in disposable cardboard or plastic returnable trays. • Intended use: For marketing as a fresh product for human consumption.
Flow diagram *(Stages 6 & 7)*	There are five key steps in the operation. These comprise site preparation, crop establishment, crop agronomy, crop harvest and post-harvest handling on the organic holding. The sequence of operations and product flow is shown in the Process Flow Diagram. The flow diagram was verified by the team as representing the most likely production options on *dd/mm/yy*.

Hazard analysis *(Stages 8 to 12)*	The hazards and controls are detailed in the Hazard Analysis Chart. In lettuce production control of hazards is dependent on reducing the likelihood of the introduction or proliferation of a hazard. No CCPs are identified. For hazards at specific steps the controls are designated operational PRPs (OpPRP), and monitoring procedures and corrective actions have been established.

Verification
(Stage 13)
	The following verification procedures are undertaken:
	• Audits of the HACCP system
	  - Internal by the company of PRPs, operational PRPs - at least annually prior to the HACCP review
	  - External by verifiers of the adopted organic assurance scheme.
	• Monitoring of customer satisfaction, including anomalies and rejections
	• Product testing, e.g. pesticide residue analysis by the producer or producer's customer or through participation in a third party monitoring system
	
	Review of the HACCP system:
	• Periodic review, e.g. annually before crop season
	• Prior to significant changes (outside the periodic review) to the process operation and hazards.

Documents and record keeping *(Stage 14)*	The following documents are retained: • The HACCP plan (including previous versions) • Management policies and operational procedures • Relevant organic scheme protocols, Codes of Practice, guidelines The following records are taken: • Operational records, e.g. site and crop management records, crop input applications • Monitoring data • Corrective actions taken • Verification data Records are retained for a minimum of three years.

† Stages in a HACCP study (See Section 1 Figure 1)

Lettuce production flow diagram

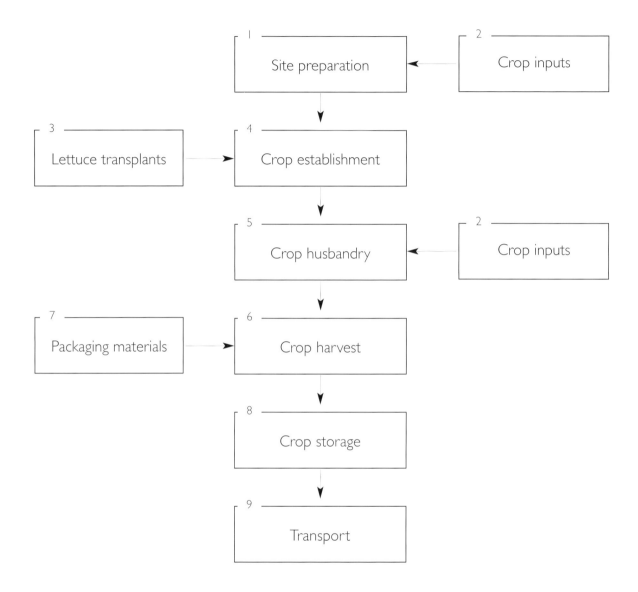

Prerequisite Programme Chart

Prerequisite programme (PRP)	Hazards controlled by the PRP	Checking procedures	Remedial actions
Personal hygiene Personal hygiene standards: - staff are made aware of the importance of their own personal hygiene, - suitable and sufficient hand washing and toilet facilities are provided, - persons with communicable enteric diseases are excluded from handling lettuce.	Introduction of food-borne pathogens from people handling lettuce.	Scheduled inspection (at an appropriate frequency) of hygiene standards, and record.	Review procedures and training needs. Record actions taken.
Hygiene and housekeeping All equipment and facilities are routinely cleaned to a defined schedule. Includes harvesting equipment, containers, vehicles, and trailers used to handle and transport lettuce.	Introduction of food-borne pathogens from equipment used for harvesting and handling lettuce.	Scheduled inspection (at an appropriate frequency) of equipment and facilities for fitness for purpose, and record.	Review procedures and take appropriate action to remedy any defects. Record actions taken.
	Contamination of organic products with non-organic crop residues and other materials.	Scheduled inspection (at an appropriate frequency) of equipment and facilities for fitness for purpose, and record.	Review procedures and take appropriate action to remedy any defects. Record actions taken.
Pest control Rodent control procedures - Inspection and treatment of premises to deter and eradicate pest ingress (rodents and birds) - Animal activity is discouraged within the crop, and livestock and domestic animals are prevented from accessing crop areas - Premises are designed and maintained to prevent entry of birds and rodents	Introduction of food-borne pathogenic bacteria from pests (birds and rodents) and animals (domestic and livestock).	Scheduled inspection (at an appropriate frequency) of facility for pest and animal activity.	Review procedures and take appropriate action to remedy any potential pest ingress. Record actions taken.

Prerequisite programme (PRP)	Hazards controlled by the PRP	Checking procedures	Remedial actions
Glass policy Glass is covered or guarded to prevent contamination of goods. Glass breakage procedures are in place.	Introduction of glass from equipment and the environment (machinery, lights, windows, etc. - field and crop handling areas).	Scheduled inspection (at an appropriate frequency) of glass fixtures and fittings.	Take appropriate action to remedy any defects. Record actions taken.
Temperature control Harvested lettuce is cooled within a set time period and subsequently stored and despatched at chill temperature (≤ 5°C).	Growth of food-borne pathogens due to time and temperature abuse.	Scheduled check (at an appropriate frequency) of chill store and despatch vehicle temperature, and time to cooling.	Review procedures and take appropriate action to remedy any defects.
Supplier assurance Raw materials (lettuce transplants, crop inputs, etc.) are sourced from reputable suppliers. Purchasing of materials takes into account the source and the treatement that it may have undergone.	Presence of food-borne pathogens, non-permitted pesticides, glass and non-organic materials due to contamination at previous stages.	Review approved supplier list. Supplier performance is evaluated seasonally.	Review suitability of suppliers.
Organic protocols The adopted organic standard is complied with in totality	Loss of organic integrity due to: - Use of inputs not authorised by organic protocols - Use of transplants from uncertified suppliers - Contamination of harvested crop with non organic crop residues from harvesting and handling equipment	Planned internal audit of procedures (at an appropriate frequency). Review of external audits by third parties and verifiers of organic schemes.	Review procedures and take necessary actions for any non-compliance identified. Record actions taken
Training Personnel are trained, instructed and supervised commensurate with their activity and competent to undertake the tasks required. Where persons are engaged in activities relating to operational PRPs and CCPs, relevant training is given.	Introduction of hazards or loss of organic integrity due to failure to follow correct procedures	Scheduled review (at an appropriate frequency) of staff competence to carry out activities.	Review training needs. Record actions taken.

Hazard Analysis Chart

Process step	Hazard	Control	CCP or Op PRP	Critical limit	Monitoring procedure	Corrective action
1. Site preparation (Selection and preparation of land for lettuce production, including any cultural operations)	No identified hazard not covered by a PRP					
2. Crop inputs (Application of crop treatments, including organic manure, fertiliser products, pesticide products and irrigation water pre and post crop establishment)	Introduction of food-borne pathogens from irrigation water	Sources of water used are identified and evaluated for suitability for intended use	Op PRP		Testing of water for microbial loading. The frequency is based on risk assessment.	Review acceptability of water source. Evaluate risk of use on crop. Record actions taken.
	Introduction of food-borne pathogens from organic manure	Permitted materials are used in the recommended manner (method and timing of application) taking into account the source and treatment that it has undergone prior to use.	Op PRP		Scheduled check of manure use records for compliance.	Review procedures. Evaluate risk to crop of possible misuse. Record actions taken.
	Introduction of pesticides from use.	Permitted pesticide products are used in the recommended manner.	Op PRP		Check pesticide product application records prior to harvest for compliance.	Review procedures. Evaluate risk to crop of possible misuse of pesticides. Record actions taken.
		Training/competence of decision makers - staff, advisors (PRP)				

Process step	Hazard	Control	CCP or Op PRP	Critical limit	Monitoring procedure	Corrective action
2. Crop inputs (contd)	Introduction of pesticides from equipment	Planned preventative maintenance and calibration of equipment. Training of operators -staff, contractors (PRP)	Op PRP		Check equipment maintenance/calibration records prior to harvest for compliance.	Review procedures. Evaluate risk to crop of possible misuse of pesticides. Record actions taken.
3. Lettuce transplants (Purchase, receipt and temporary holding of young lettuce plants)	Presence of non-organic materials from suppliers and loss of organic integrity.	Use of certified organic suppliers	Op PRP		Check organic status of supplier at intake	Review approval status of supplier Review acceptability of supplied materials
4. Crop establishment (Planting operations including any cultural operations)	No identified hazard not covered by a PRP					
5. Crop husbandry (Management of the developing crop)	No identified hazard not covered by a PRP					
6. Crop harvest (Harvesting operations in field including cutting, trimming, packing of harvested lettuce and transfer to crop store facility)	Introduction of food-borne pathogens from people (harvesting staff)	Personal hygiene standards Staff receive training in hygiene (PRP)	Op PRP		Scheduled audit of hygiene standards (procedures and facilities) and staff records (hygiene training and medical screening).	Review procedures and training needs. Record actions taken.

Process step	Hazard	Control	CCP or Op PRP	Critical limit	Monitoring procedure	Corrective action
6. Crop harvest (contd)	Introduction of pesticides due to incorrect harvest interval.	Harvest procedures - agronomist clears crop for harvest.	Op PRP		Check pesticide product application records prior to . harvest	Review procedures. Evaluate risk to crop of possible misuse of pesticides. Record actions taken.
7. Packaging (Purchase, receipt and on-site storage of packaging materials)	No identified hazard not covered by a PRP					
8. Crop storage (Chilling of packed lettuce to remove field heat and temporary storage prior to transport)	No identified hazard not covered by a PRP					
9. Transport (Loading of despatch vehicle and transport to customer at 5-8°C)	No identified hazard not covered by a PRP					

2.3 Organic apples

2.3.1 Hazards and causes in organic apple orchards

Typical hazards

Broadly speaking, hazards associated with primary organic products may be classified into three types: food safety issues, product quality attributes and organic integrity issues relating to the organic status of the product and organic production system. There are numerous theoretical food safety and product quality hazards in primary agricultural products, but a few will only be of significance in a particular agricultural situation, be this organic or conventional agriculture. In deciding the significance of a hazard, the risk associated with the hazard will have to be taken into account. This is where organic and conventional agriculture may differ; the hazard may be the same in both situations but the risk may be different.

A food safety hazard is a biological (particularly microbiological), chemical or physical agent in or on a food product with the potential to cause an adverse effect on the health of the consumer. This is very much in line with food safety legislation in Europe, which has the aim of protecting the consumer.

A product quality hazard may be defined as an attribute that causes an adverse effect on the acceptability of the product to the customer, whether this is the consumer or food business using the product. Quality attributes in fruits may relate to physical appearance (e.g. shape, size, colour, defect levels), composition including nutritional aspects, and sensory characteristics including flavour and texture.

An organic integrity hazard is defined as the product and production process not complying with the adopted organic standard and production protocol, be this set by legislation, an organic organisation such as IFOAM, or a private organic standard.

The actual hazards in any particular situation, however, will depend on the specific circumstances, including the production system, production location, product type and intended market. Many of the hazards will be the same for organic as conventionally produced products, particularly food safety issues, but the risk of the hazard may be different. However, many of the 'extended product' quality attributes which relate to the way the product is produced and the nature of the production system may be unique to organic agriculture.

In some cases the same hazard may be perceived differently in conventional and organic products. For example, the risk of the introduction of food-borne pathogens may be considered greater in organic systems. This may in part be due to a greater reliance on the use of organic manure. Conversely, the presence of pesticide residues may be perceived as greater in conventional systems due to the greater reliance on the use of pesticides in the pest, disease and weed control strategy. Therefore, the likelihood of introducing residues of pesticide residues in conventional production is high in comparison with organic products.

However, it may be deemed that the severity of the consequence of pesticide residues in organic production may be higher than in conventional products. This is due to the fact that the presence of pesticide residues at substantive levels in conventionally produced products may be acceptable (at least if below any prescribed Maximum Residue Level), whereas in organically produced products the presence of pesticide residues may be less acceptable.

Pests and diseases are difficult to control in organic orchards. An important strategy to minimise economic damage is to balance the insect pest and beneficial insect populations by providing refuges for earwigs and plant biodiversity to attract the parasitic wasps, hoverflies and lacewings. Some chemicals may be used in monitoring traps or for mating disruption, but the risk of pesticide residues arising from their use in the orchard is usually considered to be low.

Fungal diseases such as scab and mildew are typically controlled by using vigorous rootstocks on resistant varieties. Cultural controls include the eradication of overwintering infection and pruning to ensure good air circulation. In some circumstances copper and sulphur sprays are permitted in organic production to control fungal disease.

There are different organic standards for fruit production. In Europe, the EU Regulation 834/2007 is the legal minimum requirement that needs to be met for products that are designated 'organic'. Most national implementation rules are very similar to the Regulation. However, private standards may differ and impose a higher standard, including additional requirements. An example of a private standard is the Soil Association in the UK. Demeter is a private standard for biodynamic agriculture that goes beyond organic farming. Details of the differences between the EU Regulation and private standards in Europe are available on the Organic Rules website *www.organicrules.org*.

Possible causes

It is important to be aware and understand the possible cause or source of the identified hazards, as this will help determine the risk associated with the hazard and the most effective control measures. It should noted that the cause of a hazard and the level of risk may be of similar in both organic and conventional production, for example introduction of food-borne pathogens from people or equipment. However, there may be some causes that are specific to organic production, for example compromising organic integrity, or the level of risk may be perceived to be greater in organic production, for example the introduction of pathogens from organic manure.

To identify the cause of a hazard it is helpful to distinguish between those that at *present* in raw materials (e.g. inputs such as organic manure, irrigation water), those that are *introduced* during the production process (e.g. from or due to people, equipment, the environment), and changes to pre-existing hazards by *growth* or *survival* of the hazard. In crop production, presence and introduction are the predominant causes of hazards. For example in a situation where organic produce is graded at a packhouse which handles non organic produce, it may be relevant to consider the risk of cross contamination of apples with pesticide residues introduced during grading and packing.

In addition, control measures in organic production may be different to those in conventional production as there are restrictions or guidelines on what is possible within an organic system. This would not necessarily be the case for food safety issues where controls applied may be the same in organic systems as in conventional agriculture. However, for product quality and organic integrity issues controls methods would have to take account of the practices outlined in the adopted organic production standard. Many of these have an agricultural dimension and for these reference would have to be made to the adopted standard. In addition, there may be differences in the requirements in regulations, national implementation rules and private standards in the controls that are applied.

Reduction in levels of fungal and bacterial disease in organic orchards is possible by using lower tree planting densities. Conventionally managed orchards which have been established under conventional management and have been converted to organic production may need up to 50% of the trees removed to improve light penetration and quicker drying of leaves and fruit after rain or dew. (Tamm 2004).

Higher levels of flavonols and polyphenols have been found in apples from organic compared to conventional orchards. This is due to the higher level of pest and disease pressure in orchards under organic management. (Weibul 2004)

Organic apple production is most successful when the orchard environment includes hedges, wild flower strips and ecological areas which provide biodiversity of flora and fauna and enhance the population of beneficial antagonists for pest control. (Wyss 1999)

Weibul *et al* (2004) System comparison of integrated and organic apple production. Part III. Content of phenolic substances. Swiss Journal of Fruit and Wine 140 (19) 6-9

Tamm *et al* (2004) Organic fruit production in humid climates of Europe. Acta Horticulturae (ISHS) 638 333-339

Wyss (1999) Wyss *et al.* The potential of three native insect predators to control rosy apple aphid, *Dysaphis plantaginea.* Entomologica Experimentalis et Applicata 44 171-182

2.3.2 Typical hazards and the cause/source

Hazards

Hazard category	Hazard type	Hazard
Food safety	Biological	Food-borne pathogens: pathogenic bacteria (e.g. *E.coli*, *Salmonella*), viruses and protozoa
	Chemical	Pesticide residues, Mycotoxin e.g. Patulin
	Physical	Food contaminant residues (e.g. heavy metals) Foreign bodies (e.g. glass, metal, wood)
Food quality	Biological	Spoilage organisms, e.g. development of damaging fungal and bacterial rots during crop development or crop storage
	Chemical	Nutritional attributes, e.g. attainment of adequate sugar, starch, vitamin or antioxidant status
	Physical	Defects, e.g. effect of pest and disease on cosmetic appearance of fruit
		Sensory characteristics, e.g. failure to meet specification for colour, flavour and texture
Organic integrity		Organic standard requirements, e.g. <ul><li>3 year conversion period for organic apple orchard</li><li>permitted pesticides for organic crops</li><li>permitted fertiliser and manure for organic crops</li><li>wooden harvest boxes not treated with preservative</li><li>dedicated, labelled storage facility</li></ul>

Cause/source

Hazard	Cause
Food-borne pathogens	Whether introduced from • people (personal hygiene) • unclean equipment • environment (e.g. pests, stock and domestic animals) Whether present in inputs (e.g. irrigation water, organic manure)
Pesticides	Whether substances are permitted in organic systems and are used in the recommended manner Whether introduced from • inaccurate application due to people, equipment • environmental contamination (e.g. soil, adjacent crops) • equipment contamination
Glass	Whether introduced from • people • environment (e.g. lights)
Rots and moulds	
Crop variety	Whether varieties grown in parallel production are different and easily differentiated.
Site	Whether organic production takes place on clearly defined units of land such that production and post-harvest handling areas are clearly separated from non-organic production areas
Manure	Whether prohibited materials are used. Whether manure management and application compiles with the adopted organic protocol
Mineral fertilisers and supplementary plant nutrients	Whether permitted materials are used in the recommended manner.

2.3.3 Example HACCP study

[*Name of Producer*] ORGANIC APPLE ORCHARD HACCP PLAN

Last review date
(Stage 13)†   dd/mm/yy

Terms of ref. *(Stage 2)*

Objective:	This is a demonstration study covering organic integrity criteria and food safety for apples at the point of despatch of the fruit to the customer (produce marketing organisation or retailer).
Business:	*Name and location of producer*
Process:	Crop production, harvesting and storage of apples from orchard management to transport of fruit to customers.
Hazards:	**Microbiological**: Food-borne bacterial pathogens (*E.coli* and *Salmonella*)* **Chemical**: Pesticide residues above statutory MRLs *, Non-permitted pesticide products **Physical**: Glass*, Apple quality during storage, Integrity of organic produce ** Food safety hazards*
Prerequisite programmes:	The HACCP is underpinned by prerequisite programmes (PRPs) which are selected to maintain a hygienic fruit production and post-harvest handling environment. Details of the PRPs, hazards controlled by the PRP and verification procedures are given in the Prerequisite Programme Chart.
HACCP team and skills *(Stage 3)*	Farm Manager (agronomy and management skills), Agronomist (consultant with agronomy and HACCP skills), Harvest Manager (production and crop storage skills).

Essential product characteristics
(Stages 4 & 5)

- Product: Apples
- Shelf life: 8 days after packing.
- Storage and transport conditions: Storage 1-5°C. Distribution 4-7°C.
- Packaging; The crop is harvested into wooden boxes, which are also used for subsequent storage. After grading, apples are packed into polythene bags for marketing
- Intended use: For marketing as a fresh product for consumption without further preparation by the general population.

Flow diagram
(Stages 6 & 7)

There are twelve key steps in the operation. These comprise the orchard dormant period, pruning operations, crop development phase, application of crop inputs, crop harvest and post-harvest storage, grading and packing on the organic holding. The sequence of operations and product flow is shown in the Process Flow Diagram.

The flow diagram was verified by the team as representing the most likely production options on *dd/mm/yy*.

Hazard analysis
(Stages 8 to 12)

The hazards and controls are detailed in the Hazard Analysis Chart. No CCPs are identified; controls are all designated operational PRPs (OpPRP). Control of hazards is dependent on reducing the likelihood of the introduction or proliferation of a hazard. For each Operational PRP critical limits, monitoring procedures and corrective actions have been established.

Verification
(Stage 13)

The following verification procedures are undertaken:
- Audits of the HACCP system
  - Internal by the company of PRPs, operational PRPs - at least annually prior to the HACCP review
  - External by verifiers of the adopted organic assurance scheme.
- Monitoring of customer satisfaction, including anomalies and rejections
- Product testing, e.g. pesticide residue analysis
- Review of the HACCP system
  - Periodic review, e.g. annually before new crop season
  - Prior to significant changes (outside the periodic review) to the process operation and hazards.

Documents and record keeping
(Stage 14)

The following documents are retained:
- The HACCP plan (including previous versions)
- Management policies and operational procedures
- Relevant organic scheme protocols, Codes of Practice, guidelines

The following records are taken:
- Operational records, e.g. site and orchard management records, crop input applications and records of storage conditions.
- Monitoring data
- Corrective actions taken
- Verification data

Records are retained for a minimum of three years.

Apple crop flow diagram

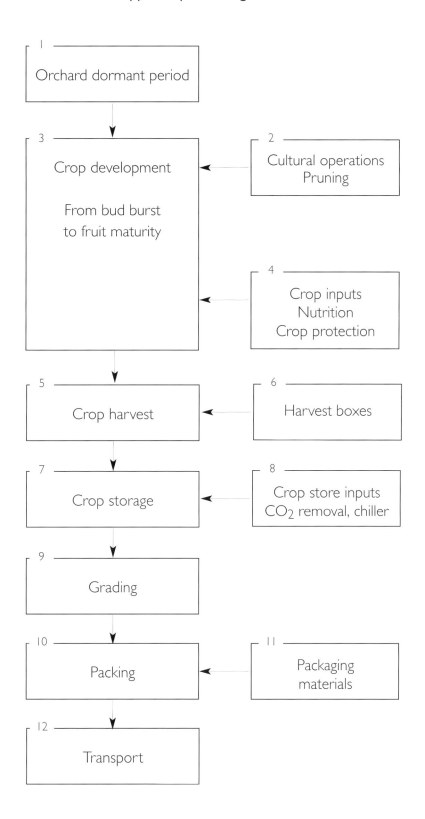

Prerequisite Programme Chart

Prerequisite programme (PRP)	Hazards controlled by the PRP	Checking procedures	Remedial actions
Personal hygiene Personal hygiene standards: - staff are made aware of the importance of their own personal hygiene, - suitable and sufficient hand washing and toilet facilities are provided, - persons with communicable enteric diseases are excluded from handling apples	Introduction of food-borne pathogens from people handling apples during harvest and grading/packing operations.	Scheduled inspection (at an appropriate frequency) of hygiene standards, and record.	Review procedures and training needs. Record actions taken.
Hygiene and housekeeping All equipment and facilities are routinely cleaned to a defined schedule. Includes harvesting and grading equipment, containers, vehicles, and trailers used to handle and transport apples.	Introduction of food-borne pathogens from equipment used for harvesting and handling apples.	Scheduled inspection (at an appropriate frequency) of equipment and facilities for fitness for purpose, and record.	Review procedures and take appropriate action to remedy any defects. Record actions taken.
	Contamination of organic products with non-organic crop residues and other materials.	Scheduled inspection (at an appropriate frequency) of equipment and facilities for fitness for purpose, and record.	Review procedures and take appropriate action to remedy any defects. Record actions taken.
Pest control Rodent control procedures - Inspection and treatment of storage and grading premises to deter and eradicate pest ingress (rodents and birds) - Animal activity is discouraged within the orchard, and livestock and domestic animals are prevented from accessing crop areas - Storage and grading premises are designed and maintained to prevent entry of birds and rodents	Introduction of food-borne pathogenic bacteria from pests (birds and rodents) and animals (domestic and livestock).	Scheduled inspection (at an appropriate frequency) of facility for pest and animal activity.	Review procedures and take appropriate action to remedy any potential pest ingress. Record actions taken.

Prerequisite programme (PRP)	Hazards controlled by the PRP	Checking procedures	Remedial actions
Glass policy Glass is covered or guarded to prevent contamination of goods. Glass breakage procedures are in place.	Introduction of glass from equipment and the environment (machinery, lights, windows, etc. - orchard and crop handling areas).	Scheduled inspection (at an appropriate frequency) of glass fixtures and fittings.	Take appropriate action to remedy any defects. Record actions taken.
Management of storage conditions Harvested apples are cooled within a set time period and a suitable storage atmosphere is established. Subsequently stored and despatched at specified temperatures.	Growth of food-borne pathogens due to time and temperature abuse.	Scheduled check (at an appropriate frequency) of chill store and despatch vehicle temperature, and time to cooling.	Review procedures and take appropriate action to remedy any defects.
	Poor apple quality following the development of storage rots and disorders resulting in scald, blotch and necrosis symptoms	Scheduled check of storage temperature and storage atmosphere	Adjust temperature and atmosphere to desired levels. Record actions
Supplier assurance Raw materials (rootstocks, crop inputs, harvest boxes, etc.) are sourced from reputable suppliers. Purchasing of materials takes into account the source and the treatment that it may have undergone.	Presence of food-borne pathogens, non-permitted pesticides, glass and non-organic materials due to contamination at previous stages.	Review approved supplier list. Supplier performance is evaluated seasonally.	Review suitability of suppliers.
Organic protocols The adopted organic standard is complied with in totality	Loss of organic integrity due to: - Use of inputs not authorised by organic protocols - Contamination of harvested crop with non organic crop residues from harvesting and handling equipment	Planned internal audit of procedures (at an appropriate frequency).	Review of external audits by third parties and verifiers of organic schemes. Review procedures and take necessary actions for any non-compliance identified. Record actions taken
Training Personnel are trained, instructed and supervised commensurate with their activity and competent to undertake the tasks required. Where persons are engaged in activities relating to operational PRPs and CCPs, relevant training is given.	Introduction of hazards or loss of organic integrity due to failure to follow correct procedures	Scheduled review (at an appropriate frequency) of staff competence to carry out activities.	Review training needs. Record actions taken.

Hazard Analysis Chart

Process step	Hazard	Control	CCP or Op PRP	Critical limit	Monitoring procedure	Corrective action
1. Dormant winter period	No specific hazard identified					
2. Cultural activities, such as pruning to develop optimal tree shape	No specific hazard identified					
3. Crop development, leaf and bud burst followed by fruit development	No specific hazard identified					
4. Application of crop inputs (Application of crop treatments, including organic manure, fertiliser products, pesticide products, mineral supplements and irrigation water.	Introduction of food-borne pathogens from irrigation water	Sources of water used are identified and evaluated for suitability for intended use	Op PRP	Water is suitable for use	Testing of water for microbial loading. The frequency is based on risk assessment.	Review acceptability of water source. Evaluate risk of use on crop. Record actions taken.
	Introduction of food-borne pathogens from organic manure	Permitted materials are used in the recommended manner (method and timing of application) taking into account the source and treatment that it has undergone prior to use.	Op PRP	Recommendations for materials and use are complied with.	Scheduled check of manure use.	Review procedures. Evaluate risk to crop of possible misuse. Record actions taken.
	Introduction of pesticide residues on harvested apples arising from scheduled use in pest and disease control	Pesticide products which are permitted in organic systems are used in the recommended manner.	Op PRP	Pesticide product recommendations are complied with.	Check pesticide product application records prior to harvest.	Review procedures. Evaluate risk to crop of possible misuse of pesticides. Record actions taken.

Process step	Hazard	Control	CCP or Op PRP	Critical limit	Monitoring procedure	Corrective action
4. Application of crop inputs (contd)		Training/competence of decision makers - staff, advisors (PRP)				
	Introduction of pesticide residues from equipment	Planned preventative maintenance and calibration of equipment.	Op PRP	Equipment is maintained and in full working order.	Check equipment maintenance/calibration records prior to harvest.	Review procedures. Evaluate risk to crop of possible misuse of pesticides. Record actions taken.
		Training of operators -staff, contractors (PRP)				
	Presence of non-organic materials from suppliers of crop inputs such as fertiliser, resulting in loss of organic integrity.	Use of certified organic suppliers	Op PRP	Only organic certified suppliers are used	Check organic status of supplier at intake	Review approval status of supplier. Review acceptability of supplied materials
5. Crop harvest (Harvesting operations in the orchard, selection of apples for packing or transfer to crop store facility)	Introduction of food-borne pathogens from people (harvesting staff)	Personal hygiene standards	Op PRP	Harvest staff comply with personal hygiene standards	Scheduled audit of hygiene standards (procedures and facilities) and staff records (hygiene training and medical screening).	Review procedures and training needs. Record actions taken.
		Staff receive training in hygiene (PRP)				
	Introduction of pesticides due to incorrect harvest interval.	Harvest procedures - agronomist clears crop for harvest.	Op PRP	Pesticide product harvest interval is adhered to.	Check pesticide product application records prior to harvest.	Review procedures. Evaluate risk to crop of possible misuse of pesticides. Record actions taken.

Process step	Hazard	Control	CCP or Op PRP	Critical limit	Monitoring procedure	Corrective action
6. Harvest boxes Wooden boxes used to transport apples from orchard and into store	Introduction of glass/ wood/metal arising from contamination of boxes	Box maintenance and cleaning policy (PRP)				
	Introduction of chemical contaminant from wood preservative	Box specification to require construction using organic approved materials (PRP)				
7. Crop storage (Removal of field heat, establishment of temperature and atmosphere for long term storage)	Development of storage rots and disorders resulting in scald, blotch and necrosis symptoms.	Apple storage quality controlled by pre requisite programme				
8. Storage inputs Establishment of correct atmosphere and temperature. Monitoring of fruit quality.	Development of storage rots and disorders resulting in scald, blotch and necrosis symptoms.	Apple storage quality controlled by pre requisite programme				
9. Grading Removal from store, size grading and selection of apples which meet quality criteria	Introduction of food borne pathogens by grading staff	Contamination hazard controlled by pre requisite programme hygiene policy				
10. Packing Apples weighed or counted into polythene bags						

Process step	Hazard	Control	CCP or Op PRP	Critical limit	Monitoring procedure	Corrective action
11 Packaging materials Supply of packaging materials (bags and boxes) of food grade quality	Introduction of food borne pathogens from contaminated packaging	Contamination hazard controlled by pre requisite programme supplier assurance policy				
12. Transport (Loading of despatch vehicle and transport to customer at 5-8°C)	Loss of shelf life resulting from failure to maintain adequate cool chain	Quality hazard controlled by pre requisite programme, temperature control				

2.4 Organic eggs

2.4.1 Hazards and causes in organic eggs

Broadly speaking, hazards associated with primary organic products may be classified into three types: food safety issues, product quality attributes and organic integrity issues relating to the organic status of the product and organic production system. There are numerous theoretical food safety and product quality hazards in primary agricultural products, but a few will only be of significance in a particular agricultural situation, be this organic or conventional agriculture. In deciding the significance of a hazard, the risk associated with the hazard will have to be taken into account.

A food safety hazard is a biological (particularly microbiological), chemical or physical agent in or on a food product with the potential to cause an adverse effect on the health of the consumer. This is very much in line with food safety legislation in Europe, which has the aim of protecting the consumer.

A product quality hazard may be defined as an attribute that causes an adverse effect on the acceptability of the product to the customer, whether this is the consumer or food business using the product. Quality attributes in eggs may relate physical appearance (e.g. shape, size, colour, defect levels), composition including nutritional aspects, and sensory characteristics including flavour and texture.

An organic integrity hazard is defined as the product and production process not complying with the adopted organic standard and production protocol, be this set by legislation, an organic organisation such as IFOAM, or an private organic standard.

The actual hazards in any particular situation, however, will depend on the specific circumstances, including the production system, production location, product type and intended market. Many of the hazards will be the same for organic as conventionally produced products, particularly food safety issues, but the risk of the hazard may be different. In some cases the risk of a hazard may be perceived differently in conventional and organic products, for example the presence of pesticide and veterinary product residues, due to the nature of the respective systems. However, many of the 'extended product' quality attributes which relate to the way the product is produced and the nature of the production system may be unique to organic agriculture.

Some examples of typical food safety and quality variables in organic eggs are shown in section 2.4.2

There are different organic standards for eggs. In Europe, the EU Regulation 834/2007 is the legal minimum requirement that needs to be met for products that are designated 'organic'. Most national implementation rules are very similar to the Regulation. However, private standards may differ and impose a higher standard including additional requirements. Examples of private standards are the Soil Association in the UK, Naturland and KAT in Germany, and KRAV in Sweden. Demeter is a private standard for biodynamic agriculture that goes beyond organic farming. Details of the differences between the EU Regulation and private standards in Europe are available on the Organic Rules website (*www.organicrules.org*).

Possible causes

It is important to be aware and understand the possible cause or source of the identified hazards, as this will help determine the risk associated with the hazard and the most effective control measures. It should noted that the cause of a hazard and the level of risk may be of similar in both organic and conventional production, for example introduction of food-borne pathogens from people or equipment. However, there may be some causes that are specific to organic production, for example compromising organic integrity, or the level of risk may be perceived to be greater in organic production, for example the introduction of pathogenic bacteria in outdoor runs from wild birds and natural food/water sources.

To identify the cause of a hazard it is helpful to distinguish between those that are *present* in raw materials (e.g. feed, new birds, housing or runs), those that are *introduced* during the production process (e.g. from or due to people, equipment, the environment), and changes to pre-existing hazards by *growth* or *survival* of the hazard.

When considering the causes of loss of product quality and organic integrity issues, it is necessary to take account of the practices outlined in the adopted organic production standard. For example, introduction of a hazard may be due to not adhering to the requirements as specified in the standard. Some possible causes apply to both organic and conventional production systems, but the risk and perception may be different. A few examples are given here.

- The outdoor run in organic farms increases the risk of introducing bacterial pathogens (*Salmonella, Campylobacter*) or chemical pollutants (dioxin, heavy metals), from wild birds, pests, atmospheric deposition or run-on of contaminated drinking water.

- In some organic farms, a substantial part of the feed is produced on-farm; this is less subject to control, resulting in a higher risk of mould and mycotoxin in the feed.

- Especially with smaller producers, selling directly to customers at the farm, the turnover may be low, resulting in longer storage periods. Longer storage periods increases the risk of old eggs or bad taste.

- On the other hand, the relatively smaller holdings of organic egg producers reduces the epidemic transmission of diseases. (The EU Organic Regulation prescribes a maximum of 3,000 birds per housing unit, but does not limit the number of units per farm; conventional egg producers can have up to 40,000 birds per housing unit).

2.4.2 Typical hazards and the cause/source

Hazards

Hazard category	Hazard type	Hazard
Food safety	Biological	Pathogenic bacteria (e.g. *E.coli, Salmonella, Campylobacter*)
	Chemical	Pesticide residues Veterinary residues Food contaminant residues (e.g. mycotoxins, heavy metals, dioxins)
Food quality	Chemical	Nutritional attributes (e.g. polyunsaturated fatty acids)
	Physical	Defects (e.g. cracks in shell, thin shells, faeces and sand residues on eggs) Old appearance (runny egg white, large air chamber) Sensory characteristics (e.g. flavour, odour and taints)
Organic integrity	Physical	Organic standard requirements, e.g. • animal welfare standards • organic origin of feed • permitted veterinary medication - number and type of treatments • identity and traceability - mixing with non-organic, mislabelling

Cause/source

Hazard	Cause

Pathogenic bacteria

Whether present in new birds, feed, water

Whether introduced from

- people (personal hygiene)
- unclean equipment, housing
- environment (e.g. wild birds, outdoor water, pests, sick/dead hens, cracked eggs)

Whether there is growth due to

- egg storage time and temperature abuse
- poor ventilation in housing

Veterinary product residues

Whether introduced by people due to

- incorrect treatments applied
- incorrect withdrawal period
- inaccurate application of treatment

Chemical contaminants (e.g. pesticides, mycotoxin)

Whether present in brought in feed (e.g. pesticides, mycotoxin)

Whether present in soil of outdoor run (e.g. dioxins)

Whether introduced in stored feed due to growth of fungal mould (mycotoxin)

Whether introduced from pest control measures (rodenticides)

Egg quality

Whether rape meal is present in feed at unacceptable level (taint)

Organic integrity

Whether there is compliance with the adopted standard, e.g.

- organic origin of new birds
- organic origin of feed
- inappropriate outdoor run

2.4.3 Example HACCP study

[*Name of Producer*] ORGANIC EGG HACCP PLAN

Last review date
(Stage 13)†   *dd/mm/yy*

Terms of ref. *(Stage 2)*

Objective:	This is a demonstration study covering food safety and product quality criteria for eggs up to the point of despatch from the farm.
Business:	*Name and location of egg producer*
Process:	Keeping, feeding and caring of hens; collecting, grading, labelling and on-farm storage of eggs, up to the point of transport from the farm.

Hazards:

The following hazards are selected as examples:
- Food safety hazards:
  - Microbiological: *Salmonella*
  - Chemical: Dioxin and Mycotoxins
- Product quality hazards:
  - Physical:
  - Dirty eggs (manure)
  - Taints (off-odour)
  - Organic integrity hazards:
  - Non-organic feed and birds
  - Residues from preventative antibiotics

Prerequisite programmes:

PRPs are identified at the outset of the HACCP study as they underpin the HACCP plan. The hazards (food safety and product quality) controlled by PRPs and procedures to verify the PRPs are presented in the Prerequisite Programme Chart.

HACCP team and skills *(Stage 3)*

The HACCP team consists of the farm manager (agronomy and management skills), the farm foreman (daily supervision of staff), and an external agronomist (consultant with food safety and zootechnical skills).

Essential product characteristics *(Stages 4 & 5)*

- Product: Graded organic eggs
- Egg storage conditions: ≤ 5°C
- Egg shelf life: 28 days between laying and consumption.
- Intended use: For marketing as a fresh product for human consumption (first grade eggs) or further processing (second grade eggs).

Flow diagram *(Stages 6 & 7)*

There are six key stages in the operation. These comprise new birds, rearing and laying, egg collection, egg grading and labelling, on-farm egg storage, and despatch and transport of eggs. An outline of all the relevant aspects of rearing of the egg laying flock and egg collection is presented in the Flow diagram. A description of the main activities with each step is given in the Hazard Analysis Chart.

The flow diagram was verified by the team as representing the most likely production operations on *dd/mm/yy*.

Hazard analysis
(Stages 8 to 12)

The hazards and controls are detailed in the Hazard Analysis Chart. In egg production control of hazards is dependent on reducing the likelihood of the introduction or proliferation of a hazard. For hazards at specific steps the controls are designated operational PRPs (OpPRP) except for significant food safety risks which are designated CCPs. Critical limits have been specified for each CCP, and monitoring procedures and corrective actions have been established for CCPs and OpPRPs.

Verification
(Stage 13)

The following verification procedures are undertaken:
- Audits of the HACCP system
  - Internal by the company of PRPs, operational PRPs and CCPs - at least annually prior to the HACCP review
  - External (six-monthly) on all hygiene aspects.
  - External by certifier of adopted organic standard.
- Monitoring of customer satisfaction, including anomalies and rejections
- Product testing, e.g.:
  - *Salmonella* analysis of eggs, hens, feed and water;
  - Pesticide residue analysis in feed and eggs;
  - Dioxin and dioxin like PCB analysis in feed and eggs (Codex: limit 3 pg dioxin and 6 pg total dioxins and dioxin-like PCBs per egg).
  - Veterinary residues in eggs.

Review of the HACCP system:
- Periodic review, e.g. annually before new flock is brought in
- Prior to significant changes (outside the periodic review) to the process operation or hazards.

Documents and record keeping
(Stage 14)

The following documents are retained:
- The HACCP plan (including previous versions)
- Management policies and operational procedures
- Relevant organic standards, Codes of Practice, guidelines (e.g. 'IKB-ei' in the Netherlands) and regulations

The following records are taken:
- Operational logbook with: all in and outputs, feeding records, health problems and veterinary treatment application records,
- Housing cleaning and maintenance records,
- Logbook of hazards occurred and corrective actions taken.
- Verification data.

Records are retained for a minimum of three years.

Egg production flow diagram

```
                              ┌─ 1 ──────────────┐
                              │  Brought in birds │
                              └──────────────────┘
                                       │
                                       ▼
┌─ 2 ──────────┐              ┌─ 4 ──────────────┐              ┌─ 5 ──────────┐
│   Housing     │ ───────────▶│ Rearing and laying│◀─────────── │     Feed      │
└──────────────┘              │                   │              └──────────────┘
                              │                   │
┌─ 3 ──────────┐             │                   │              ┌─ 6 ──────────┐
│  Outdoor run  │ ───────────▶│                   │◀─────────── │     Water     │
└──────────────┘              │                   │              └──────────────┘
                              │                   │
                              │                   │              ┌─ 7 ──────────────────┐
                              │                   │◀─────────── │ Veterinary treatments │
                              │                   │              └──────────────────────┘
┌─ 12 ─────────┐             │                   │              ┌─ 14 ─────────────┐
│ Manure storage│◀─────────── │                   │ ───────────▶│  Dead hen storage │
└──────────────┘              │                   │              └──────────────────┘
      │                       │                   │                       │
      ▼                       │                   │                       ▼
┌─ 13 ─────────┐             │                   │              ┌─ 15 ─────────────┐
│Manure disposal│             │                   │              │  Corpse disposal  │
└──────────────┘              │                   │              └──────────────────┘
                              │                   │
                              │                   │              ┌─ 16 ─────────────┐
                              │                   │ ───────────▶│ Live end-of-lay   │
                              │                   │              │ hen transport     │
                              └──────────────────┘              └──────────────────┘
                                       │
                                       ▼
                              ┌─ 8 ──────────────┐
                              │  Egg collection   │
                              └──────────────────┘
                                       │
                                       ▼
                              ┌─ 9 ──────────────┐
                              │Grading and labelling│
                              └──────────────────┘
                                       │
                                       ▼
                              ┌─ 10 ─────────────┐
                              │  Storage on farm  │
                              └──────────────────┘
                                       │
                                       ▼
                              ┌─ 11 ─────────────┐
                              │ Transport of eggs │
                              └──────────────────┘
```

Prerequisite Programme Chart

Prerequisite programme (PRP)	Hazards controlled by PRP/Result PR	Checking procedures	Remedial actions
Construction and layout of buildings and utilities: - Separation of 'clean' spaces (hens) and 'dirty' spaces (visitors, manure, corpses), with separate accesses. - 'Buffer' space for cool storage of eggs. - Paved paths and smooth floors.	Introduction of Salmonella from 'dirty' spaces or buffer spaces to clean hen housing. Introduction of Salmonella from dirty spaces or hen housing to egg storage. Survival of Salmonella on rough paths and floors.	Comparison of layout with requirements in Salmonella regulation by competent advisor; and record.	Follow recommendations for improvement. Record adjustments made.
Location and hygiene of outdoor run and fields: - Paved first part of outdoor run - Clean soil for outdoor run - No feeding outdoor - Rotating plots for outdoor run - Tree cover as shelter against predator birds - Temporary cover or temporary indoor keeping during wild bird migration season. - Clean soil in fields for growing feed crops.	Introduction of Salmonella from dirty outdoor run. Introduction of dioxins from outdoor run soil, contaminated by nearby industry. Introduction of dirt on eggs from hens in muddy outdoor run. Introduction of an organic non-compliance by inappropriate outdoor run. Introduction of pathogens from wild birds Presence of dioxins in feed grown on fields contaminated by nearby industry.	Annual analysis of soil quality in laboratory (for toxins), and record. Scheduled inspection of status of outdoor run by competent operator, and record. Participation in a Salmonella or dioxin monitoring protocol by competent operator.	Adjust location of outdoor run parcels or add clean soil. Adjust the time period spent outside. Follow recommendations for improvement. Record actions taken.
Hygiene and housekeeping - Clean housing completely before new flock. - Daily cleaning of egg collection line, grading and packing machine and floor below it. - Clean air canals - Sufficient ventilation to control temperature and humidity - Clean silos before new feed comes in - Frequent collection of eggs - Single-use trays, or recycle only clean trays from own site - Use different coded trays for pre- and post grading.	Introduction of Salmonella from previous flock. Introduction of Salmonella from dirty equipment or dirty air canals. Growth of Salmonella by too high temperature and too high humidity. Introduction of mycotoxin-producing mould from old feed. Growth of Salmonella on eggs that stay too long in housing Introduction of Salmonella from reused trays or from pre-grading trays.	Scheduled check (at appropriate frequency) on hygiene status of building and equipment by competent operator; and record in cleaning and maintenance logbook.	Review hygiene procedures and take appropriate action. Follow recommendations for improvement from third party. Record actions taken.

Prerequisite programme	Hazards controlled by PRP/Result PR	Checking procedures	Remedial actions
Supply of new birds - Buy from reputable suppliers - Buy hens with health certificate - Buy hens vaccinated against Salmonella	Presence of Salmonella on new birds	Certificate of health from supplier. Certificate of vaccination from supplier. Visit to rearing address by buyer. Record all lots of birds plus comments in operational logbook.	Analyse hens for Salmonella in lab. Destroy infected hens. Review choice of supplier. Record actions taken.
Feed supply - Buy feed from reputable suppliers - Buy feed that has been analysed	Presence of Salmonella in feed Presence of mycotoxin and forming mould in feed Presence of dioxins in feed	Certificate with results of feed analysis. Check each feed lot for visible mould by competent operator. Record all inputs plus comments in operational logbook.	Let samples be analysed. Report problem to supplier. Compost moulded feed. Review choice of supplier. Record actions taken.
Water supply and - Risk assessment of own water source. - Disinfection of water system before new flock comes in. - Clean drinking water in outdoor run and proper drainage of rainwater.	Introduction of Salmonella from contaminated water, especially in open water systems. Presence of Salmonella in drinking water outside and in runoff water.	Annual analysis of water quality in lab. Scheduled inspection of outdoor run by competent operator, and record in operational logbook.	Review choice of water source water system. Review hygiene procedures and take appropriate action. Record actions taken.
Access for visitors, suppliers and collectors: - Clear demarcation of farm boundaries. - Locked housing; doorbell; visitor register - Food supply can access silos outside housing - Egg collection from buffer room - Manure and corpses collection along roadside	Introduction of Salmonella from visitors, suppliers and collectors coming from contaminated farms. Introduction of Salmonella from vehicles coming from contaminated farms.	Regular check of entrance procedures and visitor register by competent operator, and record.	Follow recommendations for improvement. Record actions taken.
Personal hygiene: - Change clothes and shoes between 'dirty' and 'clean' spaces. - Change shoes when entering and leaving outdoor run - Disinfect shoes and wash hands when entering and leaving the hen housing	Introduction of Salmonella on staff (shoes, clothing, hands) going from dirty to clean spaces, and going from one housing unit to another.	Scheduled inspection (at appropriate frequency) of hygiene standards by competent operator, and record in cleaning and maintenance logbook.	Review hygiene procedures and take appropriate action. Record actions taken.

Prerequisite programme	Hazards controlled by PRP/Result PR	Checking procedures	Remedial actions
Training of staff (even temporary staff) to create awareness and knowledge: - Food safety - Product quality - Organic integrity	No specific hazard	Scheduled review (at appropriate frequency) of staff competence to carry out activities, by competent operator, and record.	Review training needs and undertake training programme. Record actions taken.
Control of pests and domestic animals: - Screening air inlets. - Closed housing, storage silos and waste bins. - Avoid contact with domestic animals.	Introduction of *Salmonella* by pests and domestic animals from contaminated spaces to clean housing.	Scheduled inspection (at appropriate frequency) of facilities for pest and animal activity by competent operator, and record in cleaning and maintenance logbook.	Review pest control procedures and take appropriate action. Record actions taken.
Waste management: - Daily collection of dead hens from housing to closed container, outside in cool place, inaccessible by animals, on paved ground. - Transport of corpse and manure from hen housing through 'dirty' access. - Container moved to roadside, on paved ground, just before (weekly) collection from farm. - Closed manure storage, less then 50m from hens, on paved ground.	Survival of *Salmonella* by untimely collection of dead hens. Introduction of *Salmonella* from contaminated manure and corpses due to inappropriate storage and untimely collection.	Scheduled inspection of hygiene standards of waste storage and logbook of waste collection, and record in cleaning and maintenance logbook.	Review storage facilities and waste management procedures and take appropriate action. Record actions taken.
Production according to organic standards - Outdoor run	Introduction of non-compliance with organic standards.	Regular check of compliance with organic standards by competent operator or advisor, and record.	Follow recommendations for improvement. Record actions taken

Hazard Analysis Chart: Food Safety Aspects

Process step	Hazard	Control	CCP or Op PRP	Critical limit	Monitoring procedure	Corrective action
1 New birds brought in	No identified hazard not covered by a PRP					
2 Housing	Introduction of *Salmonella* from faeces in stall and litter.	Clean stalls and put new litter before new flock is introduced.	Op PRP		Regular check of cleanliness with fresh litter and record in cleaning and maintenance logbook by competent operator.	Review cleaning procedure; review litter applications.
3 Outdoor run	Introduction of *Salmonella* from wet, muddy manure.	No accumulation of poultry manure in the outdoor run.	Op PRP		Regular check of outdoor run (e.g. after rains) and record in cleaning and maintenance logbook by competent operator.	Review choice of outdoor run. Rotate outdoor parcel with new, clean parcel, part pave outdoor run.
4 Rearing and laying	Introduction of *Salmonella* from contact with sick hens.	Take out sick or dead hens.	CCP	All sick hens are removed daily.	Check condition of birds and record in operational logbook by competent operator.	Review hygiene procedures. Destroy infected hens. Keep suspected eggs separate.
	Growth of *Salmonella* due to poor resistance of hens.	Development of resistance by feeding healthy diet (high percentage of grass and vegetables, supplement vitamins and minerals)	Op PRP		Monitor composition of diet and record in operational logbook by competent operator.	Review diet.

Hazard Analysis Chart: Food Safety Aspects

Process step	Hazard	Control	CCP or Op PRP	Critical limit	Monitoring procedure	Corrective action
4 Rearing and laying (contd)	Introduction of Salmonella from parasites	Treat parasites	Op PRP		Check for worms, and record in operational logbook.	Review parasite treatments
5 Feed	No identified hazard not covered by a PRP					
6 Water	No identified hazard not covered by a PRP					
7 Veterinary treatments	No identified hazard not covered by a PRP					
8 Egg collection	No identified hazard not covered by a PRP					
9 Grading and labelling	Introduction of Salmonella from one egg to another.	Remove dirty, cracked and broken eggs, and keep separate as second grade eggs (for industrial use) or as waste (for feed).	CCP	No cracked or too dirty eggs in storage.	Continuous visual check for cracks and dirt in stored eggs, and record problems in operational logbook.	Review visual check and selection procedure.
		Keep eggs from contaminated flock separate.	Op PRP		Scheduled check of health status of hens and storage of eggs from unhealthy hens in operational logbook.	Review health monitoring procedures. Review egg storage procedures.
	Growth of Salmonella by too long storage, due to lack of expiry date.	Correct labelling (stamping) of expiring date.	Op PRP		Regular internal control of expiry dates and record in operational logbook.	Review stamping equipment and practices. Reject eggs.

Process step	Hazard	Control	CCP or Op PRP	Critical limit	Monitoring procedure	Corrective action
10 Storage of eggs on-farm	Growth of *Salmonella* in eggs by too high temperature and too long storage.	Assure timely transport to storage (eggs are moved within 1 day to cool storage, on farm or e sewhere)	Op PRP		Check movements from / to storeroom are recorded in operational logbook	Review egg collection and storage procedures. Instruct staff.
		Assure low temperature in storage (≤ 5°C).	Op PRP		Automatic temperature checking and recording in logbook.	Review storage conditions: building and cooling equipment.
		Assure timely transport of eggs from storage (max 28 days from production to consumption, in cool storage).	Op PRP		Scheduled check of expiry date on eggs in stock and record in operational logbook by competent operator	Review egg collection and storage procedures. Reject too old eggs
11 Transport eggs	No identified hazard not covered by a PRP					
12 Manure storage	No identified hazard not covered by a PRP					
13 Manure disposal	No identified hazard not covered by a PRP					
14 Storage of dead hens	No identified hazard not covered by a PRP					
15 Corpse disposal	No identified hazard not covered by a PRP					
16 Live end-of-lay hens	No identified hazard not covered by a PRP					

Hazard Analysis Chart: Product Quality Aspects

Process step	Hazard	Control	CCP or Op PRP	Critical limit	Monitoring procedure	Corrective action
1 New birds brought in	Non-organic birds brought in	Purchase certified hens. Keep documents with operational logbook.	Op PRP		Scheduled check of records by competent operator	Replace by organic hens
2 Housing	No identified hazard not covered by a PRP					
3 Outdoor run	Dirty eggs with sand and manure	No accumulation of mud and manure in outdoor run	Op PRP		Scheduled check of outdoor run (e.g. after rains), and record in cleaning and maintenance logbook by competent operator	Review choice of outdoor run. Rotate outdoor parcel. Improve drainage.
4 Rearing and laying	No identified hazard not covered by a PRP					
5 Feed	Taints (off odour)	Avoid certain feed (such as rape seed). Record all feeding in operational log book.	Op PRP		Scheduled check of feed records in operational logbook by competent operator.	Review diet. Reject bad smelling eggs.
	Introduction of more than 15% non-organic feed in diet.	Certified feed from approved suppliers. Keep certificates with operational logbook.	Op PRP		Scheduled check of records in operational logbook by competent operator.	Review suppliers. Reject feed.
6 Water	No identified hazard not covered by a PRP					

Process step	Hazard	Control	CCP or Op PRP	Critical limit	Monitoring procedure	Corrective action
7 Veterinary treatments	Introduction of non-organic veterinary treatment from preventive antibiotics or disrespect withdrawal period after curative treatment	Use of approved products and correct doses with veterinary approval.	Op PRP		Check records of treatments in operational logbook by competent operator.	Instruct staff about standards. Decertify eggs. Decertify farm.
		Respect withdrawal period between treatment and egg collection; keep eggs separate as 'conventional'.	Op PRP		Check records of treatments in operational logbook by competent operator.	Instruct staff about standards. Decertify eggs. Decertify farm.
8 Egg collection	No identified hazard not covered by a PRP					
9 Grading and labelling	Dirty eggs with sand and manure	Dirty eggs kept separate.	Op PRP		Scheduled check of 'accepted' and 'rejected' eggs and record in operational logbook by competent operator	Review rejection procedures. Instruction of staff.
10 Egg storage on-farm	No identified hazard not covered by a PRP					
11 Transport eggs	No identified hazard not covered by a PRP					
12 Manure storage	No identified hazard not covered by a PRP					
13 Manure disposal	No identified hazard not covered by a PRP					

Process step	Hazard	Control	CCP or OpPRP	Critical limit	Monitoring procedure	Corrective action
14 Storage of slaughtered end-of-lay hens and dead hens	No identified hazard not covered by a PRP					
15 Corpse disposal	No identified hazard not covered by a PRP					
16 Live end-of-lay hens	No identified hazard not covered by a PRP					

2.5 Organic milk

2.5.1 Hazards and causes in organic milk production

Broadly speaking, hazards associated with primary organic products may be classified into three types: food safety issues, product quality attributes and organic integrity issues relating to the organic status of the product and organic production system. There are numerous theoretical food safety and product quality hazards in primary agricultural products, but a few will only be of significance in a particular agricultural situation, be this organic or conventional agriculture. In deciding the significance of a hazard, the risk associated with the hazard will have to be taken into account.

A food safety hazard is a biological (particularly microbiological), chemical or physical agent in or on a food product with the potential to cause an adverse effect on the health of the consumer. This is very much in line with food safety legislation in Europe, which has the aim of protecting the consumer.

A product quality hazard may be defined as an attribute that causes an adverse effect on the acceptability of the product to the customer, whether this is the consumer or food business using the product. Quality attributes in milk may relate the composition including nutritional aspects, and sensory characteristics including flavour and texture. In this example only the content of polyunsaturated fatty acids is considered because organic milk producers and consumers claim this nutritional value as specific for organic.

An organic integrity hazard is defined as the product and production process not complying with the adopted organic standard and production protocol, be this set by legislation, an organic organisation such as IFOAM, or an private organic standard.

The actual hazards in any particular situation, however, will depend on the specific circumstances, including the production system, production location, product type and intended market. Many of the hazards will be the same for organic as conventionally produced products, particularly food safety issues, but the risk of the hazard may be different. In some cases the risk of a hazard may be perceived differently in conventional and organic products, for example the presence of pesticide and veterinary product residues, due to the nature of the respective systems. However, many of the 'extended product' quality attributes which relate to the way the product is produced and the nature of the production system may be unique to organic agriculture.

There are different standards for organic milk. In Europe, the EU Regulation 834/2007 is the legal minimum requirement that needs to be met for products that are designated 'organic'.

Most national implementation rules are very similar to the Regulation. However, private standards may differ and impose a higher standard including additional requirements. Examples of private standards are the Soil Association in the UK, Naturland and KAT in Germany, and KRAV in Sweden. Demeter is a private standard for biodynamic agriculture that goes beyond organic farming. Details of the differences between the EU Regulation and private standards in Europe are available on the Organic Rules website (*www.organicrules.org*).

Possible causes

It is important to be aware and understand the possible cause or source of the identified hazards, as this will help determine the risk associated with the hazard and the most effective control measures. It should noted that the cause of a hazard and the level of risk may be of similar in both organic and conventional production, for example introduction of food-borne pathogens from people or equipment. However, there may be some causes that are specific to organic production, for example compromising organic integrity, or the level of risk may be perceived to be greater.

To identify the cause of a hazard it is helpful to distinguish between those that are *present* in raw materials (e.g. feed), those that are *introduced* during the production process (e.g. from or due to people, equipment, the environment), and changes to pre-existing hazards by *growth* or *survival* of the hazard. In milk production (and excluding any on farm processing) presence and introduction are the predominant causes of hazards.

When considering the causes of loss of product quality and organic integrity issues, it is necessary to take account of the practices outlined in the adopted organic production standard. For example, introduction of a hazard may be due to not adhering to the requirements as specified in the standard. Some possible causes apply to both organic and conventional production systems, but the risk and perception may be different. A few examples are given here.

- In organic farms, a substantial part of the feed is produced on-farm; this is less subject to control, which may result in a higher risk of mould and mycotoxins in the feed.

- In organic farms, the (optional) traditional straw bedding is at more risk of containing *Salmonella* than the modern housing in conventional farms.

- In organic farms, cows generally spend more time on pastures, which, especially in the summer, improves the milk quality (higher content CLA and Omega 3).

2.5.2 Typical hazards and the cause/source

Hazard category	Hazard type	Hazard
Food safety	Biological	Pathogenic bacteria: *Salmonella*, *E. coli* Disease-causing organisms: TB (*Mycobacterium tuberculosis*), Para TB, Brucellosis, Leptospirosis
	Chemical	Pesticide residues Veterinary medicine residues Food contaminant residues (e.g. mycotoxins, heavy metals, dioxins, PCBs) Cleaning chemical residues
	Physical	Glass, metal
Food quality	Chemical	Fat and protein content Unsaturated fatty acid and vitamin E content Butyric acid (cheese making)
	Physical	Wood, soil, organic matter (e.g. manure residues), insects Mineral oils (from machinery) Blood Clods from clinical Mastitis and somatic cell count from sub-clinical Mastitis Taints and off flavours and odours
Organic integrity	Physical	Organic standard requirements, e.g. • Animal welfare: < than 120 days/year outside on pasture • Straw bedding: <50% organic • Pasture: use of chemical pesticides and fertilisers • Feed: >5% non-organic in diet, <50% home grown • Veterinary medication: preventative antibiotic treatment; > 2 curative treatments per year

Cause/source

Hazard	Cause
Pathogenic bacteria	Whether present in feed and water
	Whether present in new, brought-in cows
	Whether introduced from:
	• people (personal hygiene)
	• equipment (hygiene and maintenance of milking equipment and storage tank)
	• housing (hygiene and housekeeping)
	• cow (e.g. faecal contamination of udder)
	• environment (e.g. vermin, domestic animals) and pasture (organic manure)
	Growth due to milk storage temperature and time abuse
	Survival due to pasteurisation temperature and time abuse (milk processing only)
Disease causing organisms	Whether present in new, brought-in cows
	Whether introduced from vermin, domestic animals and other stock animals
	Survival due to pasteurisation temperature and time abuse (milk processing only)
Pesticide and veterinary product residues	Whether present in feed
	Whether veterinary residues are introduced by people due to
	• incorrect treatments applied
	• incorrect withdrawal period
	• inaccurate application of treatment
Chemical contaminants	Whether present in feed (e.g. mycotoxins)
	Whether present in pasture (e.g. dioxins, heavy metals)
	Whether introduced from equipment (cleaning chemicals, mineral oils)
Physical contaminants	Whether introduced from machinery (soil and organic matter)
	Whether introduced from environment (e.g. glass lights, insects)
Milk quality	Whether introduced due to diet and feed composition:
	• taints, off odours (sugar beet, wild garlic), milk taste (roughage, aromatic herbs), colour (carrots)
	• unsaturated fatty acid and vitamin E content (roughage, clover, linseed, corn silage, concentrate feed)
	• Butyric acid content (silage)

Hazard	Cause
Milk quality (contd)	Whether introduced due to dilution with water (fat and protein content)
	Whether introduced by breed (milk taste)
	Whether introduced from Mastitis (somatic cell count (SCC), clots in milk)
Organic integrity	Whether there is compliance with the adopted standard, e.g.

- non-organic new cows
- non-organic feed
- inappropriate pasture
- preventative treatments with antibiotics
- mixing conventional and organic milk

2.5.3 Example HACCP plan

[*Name of Producer*] ORGANIC MILK PRODUCTION HACCP PLAN

Last review date
(Stage 13)†   *dd/mm/yy*

Terms of ref. *(Stage 2)*

Objective:	This is a demonstration study covering food safety and product quality criteria for milk production up to the point of collection from the farm.
Business:	*Name and location of milk producer*
Process:	Keeping, feeding and caring of cows. Milking and on-farm storage of milk up to the point of collection for onward delivery to the diary or processor.

Hazards:

The following hazards are selected as examples:
- Food safety hazards:
  - Microbiological: *Salmonella*
  - Chemical: mycotoxins
- Product quality hazards:
  - Physical:
    - Mastitis clots
    - Polyunsaturated fatty acid content
  - Organic integrity hazards:
    - Non-organic feed
    - Preventative antibiotic use

Prerequisite programmes:	PRPs are identified at the outset of the HACCP study as they underpin the HACCP plan. The hazards (food safety and product quality) controlled by PRPs and procedures to verify the PRPs are presented in the Prerequisite Programme Chart.
HACCP team and skills *(Stage 3)*	The HACCP team consists of the farm manager (agronomy and management skills), the farm foreman (daily supervision of staff), and an external advisor (consultant with food safety, HACCP and technical skills).

Essential product characteristics *(Stages 4 & 5)*

- Product: Organic milk
- Storage: 6°C if collected daily, 4°C longer term
- Intended use: For marketing as a fresh unpasteurised product to dairies and processing operations

Flow diagram
(Stages 6 & 7)

There are four key stages in the operation. These comprise new animals, animal keeping and husbandry, milking and milk storage prior to collection. An outline of all the relevant aspects of the milk production operation and related steps is presented in the Flow Diagram . A description of the main activities with each step is given in the Hazard Analysis Chart.

The flow diagram was verified by the team as representing the most likely production operations on *dd/mm/yy*.

Hazard analysis
(Stages 8 to 12)

The hazards and controls are detailed in the Hazard Analysis Chart. Controls are designated operational PRPs (OpPRP) or, for significant food safety risks CCPs. Control of hazards is dependent on reducing the likelihood of the introduction or proliferation of a hazard. For each Operational PRP and CCP critical limits, monitoring procedures and corrective actions have been established.

Verification
(Stage 13)

The following verification procedures are undertaken:
- Audits of the HACCP system
  - Internal by the company of PRPs, operational PRPs and CCPs - at least annually prior to the HACCP review
  - External (six-monthly) on all hygiene aspects.
  - External by certifier of adopted organic standard.
- Monitoring of customer satisfaction, including anomalies and rejections
- Product testing, e.g.:
  - Somatic cell count (SCC)
  - Fat and protein content
  - Veterinary residues in milk

Review of the HACCP system:
- Periodic review, e.g. annually before new stock are brought in
- Prior to significant changes (outside the periodic review) to the process operation or hazards.

Documents and record keeping
(Stage 14)

The following documents are retained:
- The HACCP plan (including previous versions)
- Management policies and operational procedures
- Relevant organic standards, Codes of Practice, guidelines and regulations

The following records are taken:
- Operational logbook with: all in and outputs, all animal movements, feeding records, health problems and veterinary treatment applications, milk storage temperature.
- Housing cleaning and maintenance records,
- Logbook of hazards occurred and corrective actions taken.
- Verification data.
Records are retained for a minimum of three years.

Organic milk production flow diagram

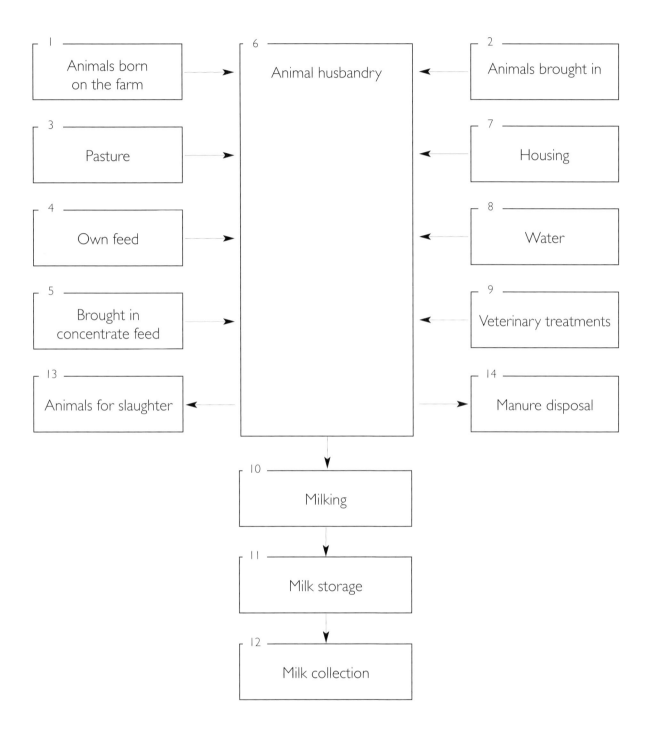

Prerequisite Programme Chart

Prerequisite programme (PRP)	Hazards controlled by PRP/Result PR	Checking procedures	Remedial actions
Construction and layout of buildings and utilities: - Separate animal holding area and milking area. - Smooth floors and paved paths, that are easily cleaned - Milk storage and manure storage separated. - Separate accesses to collect manure and milk. - Separate silos for feed requiring different moisture regimes - Separate area for animals with contagious diseases. - Preclude entrance by other domestic animals (e.g. poultry, dogs).	Introduction of *Salmonella* in milk from manure and straw. Introduction of *Salmonella* from contaminated surroundings. Introduction of *Salmonella* from manure to milk. Growth of aflatoxin-producing fungi in moist feed. Introduction of *Salmonella* from dirty vehicles to milk area. Introduction of pathogens from infected animals. Introduction of *Salmonella* from other domestic animals to milking cows	Comparison of layout with requirements in *Salmonella* regulation by competent advisor; and record.	Follow recommendations for improvement. Record adjustments made.
Hygiene of buildings: - Regular cleaning of animal holding area, especially avoiding accumulation of feed residues, mud and manure. - Regular application of fresh litter in animal holding area (but not just before milking.) - Cleaning milking area after every milking round.	Introduction of *Salmonella* from mud, manure, dirty litter, and floor.	Scheduled check (at appropriate frequency) on hygiene status of building and equipment by competent operator; and record in cleaning and maintenance logbook.	Review hygiene procedures and take appropriate action. Record actions taken.
Feed supply and storage: - Clean feed silos before refilling. - Buy feed from reputable suppliers - Keep silage feed air-tight, and other feed cool and dry. - Use oldest feed first (first in, first out)	Presence of aflatoxin or aflatoxin-producing fungi in feed.	Certificate with results of feed analysis. Check each feedlot for visible mould by competent operator; and record. Record in logbook for each feed lot: date in; temperature and humidity per week; date out.	Discard moulded feed. Report to supplier. Review choice of supplier. Review storage conditions. Record actions taken.

Prerequisite programme (PRP)	Hazards controlled by PRP/Result PR	Checking procedures	Remedial actions
Hygiene of equipment: - Clean (disinfect and rinse) milking equipment (tubes) after every milking round. - Clean milk tank after every milk collection. - Replace sieves after every milking round.	Introduction of mastitis from infected milking equipment. Introduction of Salmonella and mastitis from milk tank. Solid particles (straw and manure), which may contain Salmonella in milk.	Somatic Cell Count (SSC) of milk by milk collector.	Review hygiene procedures. Record actions taken
Hygiene of animals - Pre-treating and post-treating of teats every milking round. - Clean udders, teats, groins, flanks and abdomens of the animal. - Avoid any damage to the tissue of the teat / udder.	Introduction of mastitis and Salmonella from animal skin. Introduction of blood in milk. Introduction of Salmonella and mastitis through cow injuries. Introduction of Salmonella from faeces via dirty udders.	Scheduled check of animals for clinical mastitis, and record. Regular check of animal health by competent operator, and record.	Set mastitis animals apart. Review animal hygiene procedures. Record actions taken
Personal hygiene: - Staff and visitors use work clothing: clean clothing before each milking period. - Good personal hygiene of the milking personnel (wash hands and forearms) - Personnel carrying diseases transmittable to the milk should not enter the milking area	Introduction of pathogens to cows and milk.	Scheduled inspection (at appropriate frequency) of hygiene standards by competent operator, and record. Medical examination of milking staff.	Review hygiene procedures and take appropriate action. Record actions taken.
Milk collection - Milk collection 3 × per week, or after 5 milking rounds. - Milk haulier should wear clean clothing. - The driver should not enter the animal holding area or areas where there is manure. - The driver should change clothes and footwear if these have become dirty.	Growth of bacteria in milk to unacceptable levels. Introduction of pathogens, bacteria or other contaminants from milk haulier.	Registration of milk collection in logbook. Scheduled inspection of hygiene standards by milk haulier.	Review milk collection schedule. Review hygiene standards of milk haulier. Record actions taken.

Prerequisite programme (PRP)	Hazards controlled by PRP/Result PR	Checking procedures	Remedial actions
Access for visitors, suppliers and collectors - Access to the place of milk collection should be clear from manure, silage, etc.	Introduction of pathogens, bacteria or other contaminants from visitors.	Scheduled inspection of hygiene status of building and surroundings by competent operator, and record.	Review cleaning procedures, and take appropriate action. Record actions taken.
Water supply - Assure clean drinking water - Regular cleaning of supply pipes and trough. - Assure clean water for rinsing milking equipment.	Presence of pathogens in water from own source. Introduction of pathogens in water from dirty supply pipes and trough. Introduction of pathogens from dirty milking equipment.	Scheduled water analysis by competent operator, and record. Scheduled check of milking equipment, and record	Review water source and take appropriate action (consider tap water). Review cleaning procedures and take appropriate action. Record actions taken.
Training of staff in: - Food safety and hygienic milking procedures. - Milk quality aspects. - Organic milk production.	No specific hazard	Scheduled check (at appropriate frequency) of staff competence to carry out activities, by competent operator, and record.	Review training needs and undertake training programme. Record actions taken.
Control of domestic animals and pests - Avoid contact with other domestic animals (e.g. dogs; poultry), both in animal holding area, milking area and near feeding silos. - Proofing of housing, avoiding vermin (rodents, birds) - Regular cleaning of housing, reducing attraction of vermin - Appropriate pest control	Introduction of *Salmonella* by domestic animals, wild animals and pests	Scheduled inspection (at appropriate frequency) of facilities for pest and animal activity by competent operator, and record in cleaning and maintenance logbook.	Review animal access and pest control procedures, and take appropriate action. Record actions taken.
Waste management - Manure storage away from (preferably below) milk storage.	Introduction of *Salmonella* from stored manure to clean stable.	Compare layout with requirements in *Salmonella* regulation by competent operator, and record.	Review layout of manure storage and take appropriate action. Record actions taken.
Production according to organic standards - (example) (other aspects covered by process-step-specific Op-PRP and CCP)	Introduction of non-compliance with organic standards	Scheduled check of compliance with organic standards by competent operator or advisor, and record.	Follow recommendations for improvement. Record actions taken.

Hazard Analysis Chart: Food Safety Aspects Milk

Process step	Hazard	Control	CCP or Op PRP	Critical limit	Monitoring procedure	Corrective action
1 Animals born on farm	-					
2 Animals brought in	Presence of *Salmonella* on new bought-in animals.	Purchase from reputable suppliers.	Op PRP	No *Salmonella* on new animals	Certificate from supplier	Inform supplier. Review choice of supplier. Record actions taken
		Keep new animals in quarantine.	Op PRP	No visibly diseased animals	Check condition of new animals, and record.	Ask Vet for appropriate action. Record actions taken.
3 Pasture	-					
4 Own feed	Presence of aflatoxin in fodder (corn silage, dried grain, hay) contaminated with mycotoxin.	Use good quality fodder (separate HACCP for fodder production and harvest)	Op PRP	No visible mould, no mould smell.	Weekly inspection of feed by competent operator, and record	Discard moulded feed. Review production and harvest of feed. Record actions taken.
5 Bought concentrate feed						
6 Animal husbandry	Introduction of *Salmonella* from infected cows.	Keep suspected animals separate.	CCP	All animals with visual health problems separate.	Regular check by veterinarian. Regular check of milk (SSC)	Review animal inspection. Review veterinary treatment. Record actions taken.
7 Housing	-					
8 Water	-					

Process step	Hazard	Control	CCP or Op PRP	Critical limit	Monitoring procedure	Corrective action
9 Veterinary treatments	Antibiotic residues in milk	Respect withhold period after treatment and keep milk separate.	Op PRP	No veterinary residues.	Veterinary logbook and withhold period. Test milk on antibiotics, and record.	Review withholding procedures. Record actions taken.
10 Milking	Introduction of *Salmonella* or mastitis from personnel.	Injuries on hands or forearms should be covered with water-resistant bandage.	Op PRP	No uncovered injuries.	Scheduled check of hygiene standards of personnel, by competent operator, and record.	Review hygiene procedures. Record actions taken.
11 Milk storage	Bacterial growth due to un-cooled storage	Cool storage	CCP	Storage at 6°C when collected daily; 4°C when not collected daily.	Automatic monitoring and logbook of temperature milk tank; immediate alarm when temperature exceeds limit.	Keep milk separate. Analyse milk (SSC). Review cooling equipment. Record actions taken.
12 Milk collection	Growth of bacteria due to long storage.	Collect after max 5 milking rounds (corresponds 1-2 days)	Op PRP	5 milking rounds between 2 collections	Logbook milk collection.	Review collection schedule. Record actions taken.
	Introduction of bacteria or pathogens from suspected milk in milk factory.	Milk haulier checks the milk before collection.	CCP	No obvious visual indications of spoilage or deterioration	Report by milk haulier.	Review hygiene procedures on farm. Record actions taken.
		Milk haulier checks the storage / intake temperature before collection.	CCP	Storage time and temperature do not exceed predefined limits (6°C when collected daily; 4°C when not collected daily).	Report by milk haulier.	Keep milk separate. Analyse milk (SSC). Review cooling equipment. Record actions taken.
13 Animals for slaughter						
14 Manure disposal						

Hazard Analysis Chart: Product Quality Aspects in Milk

Process step	Hazard	Control	CCP or Op PRP	Critical limit	Monitoring procedure	Corrective action
1 Animals born on farm	-					
2 Animals brought in	-					
3 Pasture	'Introduction' of milk with low content of healthy fat	Enough other herbs besides grass.	Op PRP	Specified fat content	Visual check of pasture by competent operator, and record.	Sow herbs in pasture. Record actions taken.
	'Introduction' of non-compliance with organic, and of milk with low content in healthy fat, by animals spending insufficient time outside	Animals spend 120 days per year outside.	Op PRP	120 days	Logbook	
4 Own feed	Introduction of too much butyric acid for cheese production by low quality or too much silage	Use good quality silage feed.	Op PRP	>0.2% butyric acid in silage.	Analysis of feed by competent operator, and record.	Review silage procedures and diet. Record actions taken.
		Limit amount silage feed.	Op PRP		Logbook of feeding regime.	Review diet.
4 Own feed	Introduction of non-compliance by incorrect diet.	Feed organic feed.	Op PRP	<5% non-organic feed in diet.	Logbook of feeding regime. Scheduled check of organic feed production.	Review diet. Review feed production. Record actions taken.
5 Bought concentrate feed	Introduction of more than 5% non organic feed in diet	Feed organic feed.	Op PRP	<5% non-organic feed in diet.	Logbook of feeding regime. Certificates of bought feed.	Review diet. Review feed suppliers. Record actions taken.

Process step	Hazard	Control	CCP or Op PRP	Critical limit	Monitoring procedure	Corrective action
6 Rearing	-					
7 Housing	-					
8 Water	-					
9 Veterinary treatments	Introduction of non-compliance with organic standard	No preventative antibiotic treatment	Op PRP	No preventative treatments.	Logbook with veterinary treatments	Review treatments. Keep animals and milk separate as 'non-organic'. Record actions taken.
		Limited curative antibiotic treatment.	Op PRP	Maximum 2 curative treatments per year.	Logbook with veterinary treatments	Review treatments. Keep animals and milk separate as 'non-organic'. Record actions taken.
10 Milking	Introduction of mastitis-causing bacteria from infected udders	Visual check of animal health before each milking round.	CCP	No milk from unhealthy animals in main tank.	Scheduled check of animal health by competent person, and record. Analyse milk of suspected animals (SCC < 250,000 per ml).	Review animal inspection procedures. Keep suspected animals separate. Record actions taken.
		Check udders for infection before each milking round.	CCP	No milk from feverish hot and red udders in main tank.	Scheduled check of animal health by competent person, and record. Analyse milk of suspected animals (SCC < 250,000 per ml).	Review udder checking procedures. Keep suspected animals separate. Record actions taken.

Process step	Hazard	Control	CCP or Op PRP	Critical limit	Monitoring procedure	Corrective action
10 Milking (contd)		Visual check of foremilk before each milking round.	CCP	No milk with obvious mastitis clods in main tank.	Scheduled check of animal health by competent person, and record. Analyse milk of suspected animals (SCC < 250,000 per ml).	Review foremilk checking procedures. Keep suspected animals separate. Record actions taken.
		Keep milk from suspect animals separate or milk last and keep milk separate.	Op PRP	No milk from suspected animals in main tank.	Scheduled check of animal health by competent person, and record. Analyse milk of suspected animals (SCC < 250,000 per ml).	Review suspected animals and milk separation procedures. Record actions taken.
11 Milk storage	-					
12 Milk collection	-					
13 Animals for slaughter	-					
14 Manure disposal	-					

2.6 Organic pork meat

2.6.1 Hazards and causes in organic pork meat production

Broadly speaking, hazards associated with primary organic products may be classified into three types: food safety issues, product quality attributes and organic integrity issues relating to the organic status of the product and organic production system. There are numerous theoretical food safety and product quality hazards in primary agricultural products, but a few will only be of significance in a particular agricultural situation, be this organic or conventional agriculture. In deciding the significance of a hazard, the risk associated with the hazard will have to be taken into account.

A food safety hazard is a biological (particularly microbiological), chemical or physical agent in or on a food product with the potential to cause an adverse effect on the health of the consumer. This is very much in line with food safety legislation in Europe, which has the aim of protecting the consumer.

A product quality hazard may be defined as an attribute that causes an adverse effect on the acceptability of the product to the customer, whether this is the consumer or food business using the product. Quality attributes in pigs for pork meat may relate the physical condition of the pig before slaughter and carcass, composition including nutritional aspects, and sensory characteristics including flavour and texture. In this manual only the intramuscular fat content is considered as this is perceived to be linked to taste.

An organic integrity hazard is defined as the product and production process not complying with the adopted organic standard and production protocol, be this set by legislation, an organic organisation such as IFOAM, or an private organic standard.

The actual hazards in any particular situation, however, will depend on the specific circumstances, including the production system, production location, product type and intended market. Many of the hazards will be the same for organic as conventionally produced products, particularly food safety issues, but the risk of the hazard may be different. In some cases the risk of a hazard may be perceived differently in conventional and organic products, for example the presence of pesticide and veterinary product residues, due to the nature of the respective systems. However, many of the 'extended product' quality attributes which relate to the way the product is produced and the nature of the production system may be unique to organic agriculture.

There are different organic standards for pigs. In Europe, the EU Regulation 834/2007 is the legal minimum requirement that needs to be met for products that are designated 'organic'.

Most national implementation rules are very similar to the Regulation. However, private standards may differ and impose a higher standard including additional requirements. Examples of private standards are the Soil Association in the UK, Naturland and KAT in Germany, and KRAV in Sweden. Demeter is a private standard for biodynamic agriculture that goes beyond organic farming. Details of the differences between the EU Regulation and private standards in Europe are available on the Organic Rules website (*www.organicrules.org*).

Possible causes

It is important to be aware and understand the possible cause or source of the identified hazards, as this will help determine the risk associated with the hazard and the most effective control measures. It should noted that the cause of a hazard and the level of risk may be of similar in both organic and conventional production, for example introduction of food-borne pathogens from people or equipment. However, there may be some causes that are specific to organic production, for example compromising organic integrity, or the level of risk may be perceived to be greater.

To identify the cause of a hazard it is helpful to distinguish between those that at *present* in raw materials (e.g. feed, housing or runs), those that are *introduced* during the production process (e.g. from or due to people, equipment, the environment), and changes to pre-existing hazards by *growth* or *survival* of the hazard. In pig rearing presence and introduction may be the predominant causes of hazards.

When considering the causes of loss of product quality and organic integrity issues, it is necessary to take account of the practices outlined in the adopted organic production standard. For example, introduction of a hazard may be due to not adhering to the requirements as specified in the standard. Some possible causes apply to both organic and conventional production systems, but the risk and perception may be different. A few examples are given here.

In organic farms, a substantial part of the feed is produced on-farm, and is less subject to control, which may result in a higher risk of mould and mycotoxins in the feed.

In organic farms, the (optional) traditional housing is at more risk of containing *Salmonella* than the modern housing in conventional farms.

2.6.2 Typical hazards and the cause/source

Hazard

Hazard category	Hazard type	Hazard
Food safety	Biological	Pathogenic bacteria: *Salmonella* Parasites: Toxoplasma, round worms, Coccidia
	Chemical	Pesticide residues Veterinary residues: antibiotics Food contaminant residues (e.g. mycotoxins, heavy metals, dioxins, PCBs)
	Physical	Glass Metal
Food quality	Chemical	Water content
	Physical	Intramuscular fat content Iaints
Organic integrity	Physical	Organic standard requirements, e.g. • access to pasture • inappropriate outdoor run † • feed > 5% non-organic, <50% home grown • permitted veterinary medication - number and type of treatments

† Animal treatments that are not in line with organic principles, e.g. inappropriate outdoor run that is too small, and the presence of manure.

Cause/source

Hazard	Cause
Pathogenic bacteria and parasites	Whether present in feed and water Whether introduced from • people (personal hygiene) • unclean housing and outdoor run • environment (e.g. vermin, domestic animals)
Veterinary product residues	Whether present in feed Whether introduced by people due to • incorrect treatments applied • incorrect withdrawal period • inaccurate application of treatment
Chemical contaminants	Whether present in feed (e.g. pesticides, mycotoxins) Whether present in soil of outdoor run (e.g. dioxins)
Meat quality (low in intramuscular fat)	Whether diet is low in grain legumes
Organic integrity	Whether there is compliance with the adopted standard, e.g. • non-organic new piglets and sows • non-organic feed • inappropriate outdoor run • preventative treatments with antibiotics

2.6.3 Example HACCP study

[*Name of Producer*] ORGANIC PIG REARING HACCP PLAN

Last review date
(Stage 13)†   *dd/mm/yy*

Terms of ref. *(Stage 2)*

Objective:	This is a demonstration study covering food safety and product quality criteria for pigs reared for pork meat up to the point of despatch from the farm.
Business:	*Name and location of pig producer*
Process:	The organic pork system is 'closed': it contains both breeding sows and the fattening of pigs.
Hazards:	The following hazards are selected as examples:

- Food safety hazards:
  - Microbiological: Toxoplasma; Endo-paraistes (round worms and Coccidia)
- Product quality hazards:
  - Physical and sensory hazards:
  - Low content of intramuscular fat
  - Organic integrity hazards:
  - Insufficient access to pasture
  - Insufficient access to outdoor run

Prerequisite programmes:	PRPs are identified at the outset of the HACCP study as they underpin the HACCP plan. The hazards (food safety and product quality) controlled by PRPs and procedures to verify the PRPs are presented in the Prerequisite Programme Chart.
HACCP team and skills *(Stage 3)*	The HACCP team consists of the farm manager (agronomy and management skills), the farm foreman (daily supervision of staff), and an external advisor (consultant with food safety, HACCP and technical skills).
Essential product characteristics *(Stages 4 & 5)*	• Product: Organic pigs reared for pork meat. • Intended use: For marketing as organic pigs for fresh pork meat for human consumption or further processing (pork meat products).
Flow diagram *(Stages 6 & 7)*	There are three key stages in the operation. These comprise animals born on farm, pig husbandry (keeping breeding sows and fattening of pigs) and preparation of fattened pigs for transport to the abattoir. An outline of all the relevant aspects of the pig rearing operation and related steps is presented in the Flow Diagram . A description of the main activities with each step is given in Hazard Analysis Chart.

The flow diagram was verified by the team as representing the most likely production operations on *dd/mm/yy*.

Hazard analysis
(Stages 8 to 12)

The hazards and controls are detailed in the Hazard Analysis Chart. In egg production, control of hazards is dependent on reducing the likelihood of the introduction or proliferation of a hazard. For hazards at specific steps the controls are designated operational PRPs (OpPRP), except for significant food safety risks which are designated CCPs. Critical limits have been specified for each CCP, and monitoring procedures and corrective actions have been established for CCPs and OpPRPs.

Verification
(Stage 13)

The following verification procedures are undertaken:
- Audits of the HACCP system
  - Internal by the company of PRPs, operational PRPs and CCPs - at least annually prior to the HACCP review
  - External (six-monthly) on all hygiene aspects.
  - External by certifier of adopted organic standard.
- Monitoring of customer satisfaction, including anomalies and rejections
- Product testing, e.g.:
  - Meat and fat content
  - Intramuscular fat content
  - Veterinary residues in meat

Review of the HACCP system:
- Periodic review, e.g. annually before new stock is transferred to the fattening unit
- Prior to significant changes (outside the periodic review) to the process operation or hazards.

Documents and record keeping
(Stage 14)

The following documents are retained:
- The HACCP plan (including previous versions)
- Management policies and operational procedures
- Relevant organic standards, Codes of Practice, guidelines and regulations

The following records are taken:
- Operational logbook with: all in and outputs, all animal movements, feeding records, health problems and veterinary treatment applications,
- Housing cleaning and maintenance records,
- Logbook of hazards occurred and corrective actions taken.
- Verification data.

Records are retained for a minimum of three years.

† Stages in HACCP study (See Table 1)

Organic pig rearing flow diagram

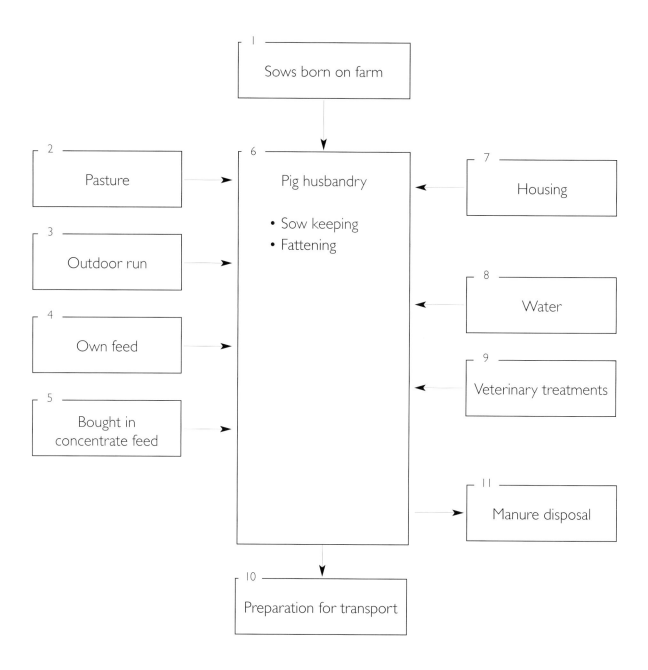

1	Sows born on farm
2	Pasture
3	Outdoor run
4	Own feed
5	Bought in concentrate feed
6	Pig husbandry • Sow keeping • Fattening
7	Housing
8	Water
9	Veterinary treatments
11	Manure disposal
10	Preparation for transport

Prerequisite Programme Chart

Prerequisite programme	Hazards controlled by PRP/Result PR	Checking procedures	Remedial actions
Construction and layout of buildings and utilities: - Smooth floors and paved paths, and easy to clean - Separate accesses to collect manure. - Outdoor run sufficiently accessible	Introduction of pathogens from contaminated surroundings. Introduction of pathogens from vehicles collecting manure. Non-compliance with organic standard	Comparison of layout with requirements (in e.g. *Salmonella* regulation), and with requirements in organic standard, by competent advisor, and record.	Follow recommendations for improvement. Record adjustments made.
Hygiene of buildings: - Regular cleaning of animal holding area, especially avoiding accumulation of feed residues, mud and manure.	Introduction of endo-parasites from mud, manure, dirty litter, and floor.	Scheduled check (at appropriate frequency) on hygiene status of building and equipment by competent operator, and record in cleaning and maintenance logbook.	Review hygiene procedures and take appropriate action. Record actions taken.
Water supply - Assure clean drinking water - Regular cleaning of supply pipes and trough.	Introduction of toxoplasma and endo-parasites in water from dirty supply pipes and trough, from faeces of cats or dead rodents.	Scheduled water analysis by competent operator, and record.	Review cleaning procedures and take appropriate action. Record actions taken.
Training of staff in: - Food safety aspects. - Pork quality aspects. - Organic pork production.	No specific hazard	Scheduled check (at appropriate frequency) of staff competence to carry out activities, by competent operator, and record.	Review training needs and undertake training programme. Record actions taken.
Control of domestic animals and pests - Avoid contact with other domestic animals (especially cats). - Proofing of housing avoiding vermin (rodents) - Regular cleaning of housing, reducing attraction of vermin - Appropriate pest control.	Introduction of toxoplasma by domestic animals, wild animals and pests.	Scheduled inspection (at appropriate frequency) of facilities for pest and animal activity by competent operator, and record in cleaning and maintenance logbook.	Review animal access and pest control procedures, and take appropriate action. Record actions taken.

Prerequisite programme	Hazards controlled by PRP/Result PR	Checking procedures	Remedial actions
Waste management - Manure storage away from housing.	Introduction of endo-parasites from stored manure to clean stable.	Comparison of layout with requirements in *Salmonella* regulation by competent operator; and record.	Review layout manure storage and take appropriate action. Record actions taken.
Production according to organic standards (other aspects covered by process-step-specific Op-PRP and CCP)	Introduction of non-compliance with organic standards	Scheduled check of compliance with organic standards by competent operator or advisor; and record.	Follow recommendations for improvement. Record actions taken.

Hazard Analysis Chart: Food Safety Aspects

Process step	Hazard	Control	CCP or Op PRP	Critical limit	Monitoring procedure	Corrective action
1 Sows born on farm	-					
2 Pasture	Introduction of endo-parasites from manure.	Treat animals after checking manure	Op PRP	Worm count (species specific threshold)	Regular check on pig health, by competent operator, and record	Review parasite treatment procedures. Review pasture. Record actions taken.
3 Outdoor run	Introduction of endo-parasites from manure.	Treat animals after checking manure	Op PRP	Worm count (species specific threshold)	Regular check on pig health, by competent operator, and record	Review parasite treatment procedures. Review outdoor run. Record actions taken.
4 Own feed	-					
5 Bought concentrate feed	-					
6 Pig husbandry	-					
7 Housing	Introduction with of toxoplasma from pigs biting ears and tails	Sufficient access to pasture and outdoor run	Op PRP		Regular check of pig behaviour and injuries, and record.	Review access to outdoor run. Record actions taken.
	Introduction of endo-parasites from manure.	Treat animals after checking manure	Op PRP	Worm count (species specific threshold)	Regular check on pig health, by competent operator, and record	Review parasite treatment procedures. Record actions taken.
8 Water	-					
9 Veterinary treatments	-					
10 Preparation for transport	-					
11 Manure disposal	-					

Hazard Analysis Chart: Quality Aspects

Process step	Hazard	Control	CCP or Op PRP	Critical limit	Monitoring procedure	Corrective action
1 Animals born on farm	-					
2 Pasture	1. 'Introduction' of non-compliance with organic, by animals having insufficient access to pasture	1.1. Animals have permanent (when not freezing) access to clean pasture.	Op PRP		Scheduled check of status of pasture and use by pigs, and record.	Review choice of pasture. Consider rotation. Record actions taken.
3 Outdoor run	1. 'Introduction' of non-compliance with organic, by animals having insufficient access to outdoor run	1.1 Arimals have access to clean outdoor run.	Op PRP		Scheduled check of status of outdoor run and use by pigs, and record.	Review cleaning of outdoor run. Record actions taken.
4 Own feed	1. Introduction of low content of intramuscular fat due to diet low in grain legumes	1.1 Use diet with sufficient grain legumes	Op PRP		Feed logbook. Feedback from butcher about pork quality.	Review diet. Record actions taken.
5 Bought concentrate feed	1. Introduction of low content of intramuscular fat due to diet low in grain legumes	1.1 Use diet with sufficient grain legumes	Op PRP		Feed logbook. Feedback from butcher about pork quality.	Review diet. Record actions taken.
6 Pig husbandry	-					
7 Housing	-					
8 Water	-					
9 Veterinary treatments	-					
10 Preparation for transport	-					
11 Manure disposal	-					